Implementing SAP SuccessFactors

A Client Centered Approach

Permanand Singh

TriMundo
Fort Myers, FL
USA
Phone: 1 (239) 247-1606

Cover Photo by Heide Randall

Published by TriMundo 03/15/2017

ISBN: 978-0-9987306-2-2 (sc)
ISBN: 978-0-9987306-0-8 (hc)
ISBN: 978-0-9987306-1-5 (e)

Contents

Foreword

to

"Implementing SAP SuccessFactors:
a client centered approach"

Permanand Singh and I first met in 2008 when we were onsite working with a SuccessFactors customer. This was the first time we worked together, but I already knew Permanand from his reputation as one of the most effective implementation consultants in the organization. As a colleague put it, "my confidence that an implementation will be successful immediately increases when I learn Permanand is on the project". Permanand's skills have continued to grow since that meeting many years ago. I suspect he is now among the top 10 most knowledgeable experts in the world when it comes to configuration, use, and capabilities of SuccessFactors technology. So I was delighted to learn he had written a book sharing his experience and insight into the effective use of SuccessFactors solutions.

Much of my professional life has focused on understanding how technology can enable human capital management (HCM) processes to effectively increase workforce productivity and support business strategies. One thing I have learned from this experience is there are no "best practices" when it comes to using HCM technology.

Methods that work well in one company may fail in another. But successful companies do tend to have some things in common. Foremost is a solid understanding of the HCM technology they are using. Technology enables and constrain process capabilities. Understanding how HCM technology works is fundamental to building HCM processes that effectively leverage it. The book "Implementing SAP SuccessFactors: a client centered approach" helps people using SuccessFactors solutions to gain this sort of in-depth understanding.

The book provides a summary review of the entire suite of SuccessFactors solutions that is accessible yet comprehensive. It walks through key steps impacting solution configuration and deployment, making a point to call out critical areas that are frequently overlooked but that significantly impact the ultimate value of the technology. A variety of topics such as project staffing, system integration, and post-deployment support are addressed in a practical and insightful manner. The book also demystifies jargon that can cause confusion for people who are new to SuccessFactors technology.

Much of the beauty of this book lies in its targeted approach. By focusing specifically on SuccessFactors technology, Permanand provides a wealth of detailed, readily applicable information for users of SuccessFactors solutions. In many ways, this book is a perfect complement to my own book "Commonsense Talent Management". Commonsense Talent Management provides guidance on designing and using HCM processes to improve organizational performance. Implementing SAP SuccessFactors provides guidance on how to use SuccessFactors technology to deploy, support and scale these HCM processes across an organization.

In an ideal world, every SuccessFactors customer and partner would receive the book Commonsense Talent Management and the book Implementing SAP SuccessFactors prior to working with SuccessFactors technology. But if you can only get one book and your primary goal is understanding the details that go into a successful deployment of SuccessFactors solutions then I heartily recommend Implementing SAP SuccessFactors as the better of the two books to read. And it is no small praise when an author recommends someone else's book over their own!

Steven Hunt, Ph.D.
Senior Vice President, Human Capital Management
SAP SuccessFactors

Acknowledgments

I have received helpful suggestions and feedback on this book from a number of people, including several of my colleagues at 3D Results. In particular, Bob Rook, CEO, and Brian Fieser, president, have provided their unique insights and generously allowed me to use content from the 3D Results implementation practices. Thanks also to the members of our marketing team—Marilyn Pearson Hendricks, Rebecca Hirschfield, and Jasmin Miah—who contributed to the development and publishing of this book in so many ways. They were instrumental to the creation of many of the graphics and assisted with proofreading, editing, and promotion.

A special thanks also to the 3DR Practice leaders who contributed content and provided their feedback. I was fortunate to have Lance Brolin, Stacey Hemiller, Heather Cook, and Kelly Rasmussen from our recruiting practice assist with the recruiting content. On the Learning, Management, and Jam modules, Betsie Reynolds provided valuable insight. Rob Comesanas did the same for the compensation content. For the Employee Central module, Lori Marra was instrumental in crafting the content for this overview, and finally, John Garten shared his experience on the core talent modules such as Goals, Performance, Succession, CDP, and 360.

Introduction

In the past few years, I have written several blog posts and documents about SAP SuccessFactors. These were primarily for experienced implementation consultants and clients that were already using the product. Similarly, a good deal of the available content around SAP SuccessFactors is targeted to experienced implementation consultants and current users. I realized there was a lack of material targeted toward newer clients and consultants, so I decided to focus this book on clients and new consultants who are just starting the selection or implementation process for a cloud-based human capital management (HCM) solution like SAP SuccessFactors.

In deciding what to include in this book, I have considered both my experience as a client and also as an implementation consultant. To that end I have taken a simple and practical approach to describing the SAP SuccessFactors HCM suite and the implementation journey. At too many implementation project kickoffs, I have seen how overwhelmed clients can get with all the information that is thrust at them.

Instead of just focusing on the implementation, I have also provided some background on the SAP SuccessFactors HCM suite. My thought here was that it is better to have some details about the systems before diving into the implementation. For the same reason, there is also a section on the "why," or the business reasons for the implementation. I have also attempted to provide more coverage on critical parts of the implementation process that are often overlooked.

In addition, I have included sections on the post-implementation period and enhancement cycles. Often, clients think that all their requirements (current and future) must be addressed in the first implementation phase. This can lead to additional complexity and lower user adoption. Focusing on key business drivers and user adoption during the design and configuration process will result in a much smoother rollout.

An SAP SuccessFactors implementation is a large project, but it brings with it the promise of an integrated human capital management system that improves both business alignment and people performance. Because of the benefits that can be realized and the resources required for the implementation, it is my hope that this book can be used as a practical introduction to the SAP SuccessFactors HCM product suite and help ensure a successful implementation for you and your organization.

Chapter 1

Overview of SAP SuccessFactors

SuccessFactors was created in 2001 and quickly became a leading vendor for cloud-based (also known as Software as a Service, or SaaS) talent management solutions. Its initial module focused on Performance and Goals Management. At the time, I was in the Information Technology department at Gartner and looking for a best of breed vendor for performance and goals. My team looked at all the vendors supporting the HR domain and could not find a vendor that met our requirements. SuccessFactors was one of the few vendors we reviewed again the following year. That time we selected them because they had already addressed some of the gaps in the product and supported more of the key features that were needed by Gartner.

After its initial focus on performance and goals, SuccessFactors rounded out the Talent Management suite with Succession, Career and Development Planning, 360 Multi-Rater, and Compensation and Variable Pay. The Recruiting Management module was developed next and then finally Employee Central, a human resources information system (HRIS).

Workforce Analytics, Learning, Recruiting, Marketing, and Onboarding were added when SuccessFactors acquired these modules from other companies. The Workforce Analytics and Planning module was added to the product suite with the Inform acquisition in 2010. In the same year SuccessFactors also acquired CubeTree. The social enterprise software created by this company is now known as the Jam module. The next year, 2011, SuccessFactors acquired Plateau, which resulted in the Learning module being added to the HCM suite. The Recruiting Marketing module was also added in 2011 through the acquisition of Job2web. In 2011 SuccessFactors itself was acquired by SAP. Two thousand and thirteen saw the addition of the Onboarding Module. This was done through the acquisition of a company called KMS Software.

Sometimes Calibration and Stack Ranker are considered modules, but since they are dependent on the other modules, I have not listed them as independent products. And for those familiar with the Job Description Manager (JDM), I have excluded that as a module since those capabilities are now in the Job Profile Builder which I consider part of the platform components.

System Architecture

SAP SuccessFactors is often described as the leader in cloud-based HCM software for talent management, core human resources (HR), and HR analytics. But what exactly is SAP SuccessFactors?

In a nutshell, SAP SuccessFactors is a cloud-based suite of products that supports all the various HR operations and functional areas. I'll provide an overview of what is cloud-based in this chapter and then delve into the different modules in this HCM suite of products in succeeding chapters.

Cloud-based means that there is no need for a client to install any hardware or implement any software or middleware. All of the system code and databases are stored in various data centers across the globe. SAP SuccessFactors uses two primary data centers in the United States. In addition, there are others in Amsterdam and other parts of the globe. As such, if your organization is subject to specific privacy laws that require employee data to be stored on a particular continent, this can be supported. Refer to Figure 1.1 for a high-level layout of the SAP SuccessFactors system architecture.

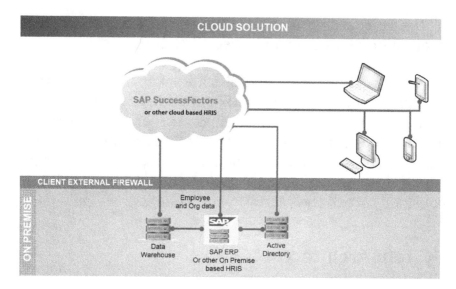

Figure 1.1: Cloud and On-Premise System Architecture

The top of the image shows the cloud-based SAP SuccessFactors system, and the lower section shows the on-premise systems behind the client's external firewall. These on-premise solutions are typically legacy applications, such as enterprise resource planning (ERP) and other custom-developed applications. There will likely be quite a few other systems and servers that are on-premise or located in your organization's data center, but for the purpose of this illustration, I am showing the ones typically affected by a talent management implementation.

The SAP SuccessFactors system resides outside of the organization's firewall, so it can be accessed by anyone with internet connectivity. Through the use of single sign-on (SSO) technology, you can still require the user to connect to your organization's network first where they will be authenticated before they can access the system. In addition, there are several layers of security built in, so you can control module and data access using role-based permissions. These will be covered in more detail by the implementation consultants when designing the system.

Before wrapping up this brief overview of the system architecture, I'll define an *instance*. This is a term you will hear often during the implementation. Some consultants will use the word *tenant* instead of *instance*. Both terms refer to the system environment that was created for your organization. For on-premise systems, there are typically development, quality assurance (or test), and production environments. For most SAP SuccessFactors implementations there will be two instances or environments—test and production. Employee Central clients may have more than two instances as an additional instance is required for parallel testing. Clients can purchase additional instances as needed from SAP SuccessFactors.

The SAP SuccessFactors HCM Suite

The actual SAP SuccessFactors HCM suite is comprised of several modules. Some are focused on talent acquisition, others on talent management, analytics and core HR, but the beauty is that they are all integrated. It is not like in the past, where to implement a best of breed HCM suite you had to cobble together multiple vendors and systems. In this case, all the back-end integration of key user data is already addressed by the vendor.

Figure 1.2 is a graphical illustration of the SAP SuccessFactors HCM suite. Employee Profile and Employee Central are located in the center because employee data is stored in these modules and made accessible to all the other modules. If Employee Central is not implemented, then Employee Profile is the repository for all employee data. If Employee Central is implemented, then it becomes the repository for employee data, and there are synchronization jobs that keep the Employee Profile and Employee Central in sync. This behind the scene synchronization is transparent to the other modules as they will continue to get updated employee data from the Employee Profile.

Figure 1.2: The SuccessFactors HCM Suite

In the following sections, I will briefly introduce each module. Since I am focusing on the overall implementation process, it is beyond the scope of this book to delve into the details of each module. For the modules overview, we are looking at the capabilities of the system, so I will review each module from a functional perspective instead of how they are licensed with SAP. This is the reason calibration is not listed as a separate module since it cannot exist independently. It needs to be implemented with Performance or Succession/Live Profile or Compensation.

I have also not included CPM (Continuous Performance Management) that was rolled out in late 2016 as this module is primarily focused on supporting the Goals and Performance processes. It does allow you to track activities you work on, links activities to business and development goals and is integrated with Performance, Goals, & Calibration. On the other hand, I have included 360 Multi-Rater as a module, as its capabilities are unique and can exist independently of the other modules.

Performance Management and Goal Management (PM/GM)

These are essentially two separate modules, but they are often implemented together. The Goal Management (GM) module is used for creating, editing, and managing performance goals. This is one of the more mature modules in the HCM suite with comprehensive goal management capabilities such as goal cascading and linking features, which can be used to ensure goal alignment across the organization. There are other features such the SMART Goal wizard and goal libraries that make it easier to create new goals. A robust security model allows the granular control of which fields can be viewed or edited. In addition, you can also create goals as either public or private goals and restrict access to the private goals to a different subset of users than the public goals.

The Goal Management module is one of the easier modules to implement, unless some of the advanced goal execution features such as status reports, execution maps, and the advanced permission features that support locking of the goal plans are implemented. Before we look at some of these features in more detail, let's first look at a sample goal plan. Figure 1.3 has a screenshot of the main page of the Goal Management module.

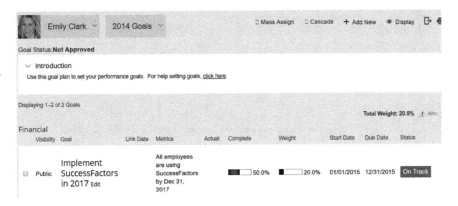

Figure 1.3: Summary View of a Goal Plan

The definition of the goal plan is based on an online worksheet that is called a goal plan template. You can have multiple goal plan templates, though the recommendation is to only have one template per year. Each year prior to rolling out your goal planning process, the goal plan template can be updated based on feedback or new requirements needed for the coming year. Your implementation consultant will help you design and configure the goal plan template based on the features you need. This configuration will impact what you can see on the main page and also when looking at the goal details as shown in Figure 1.4 below.

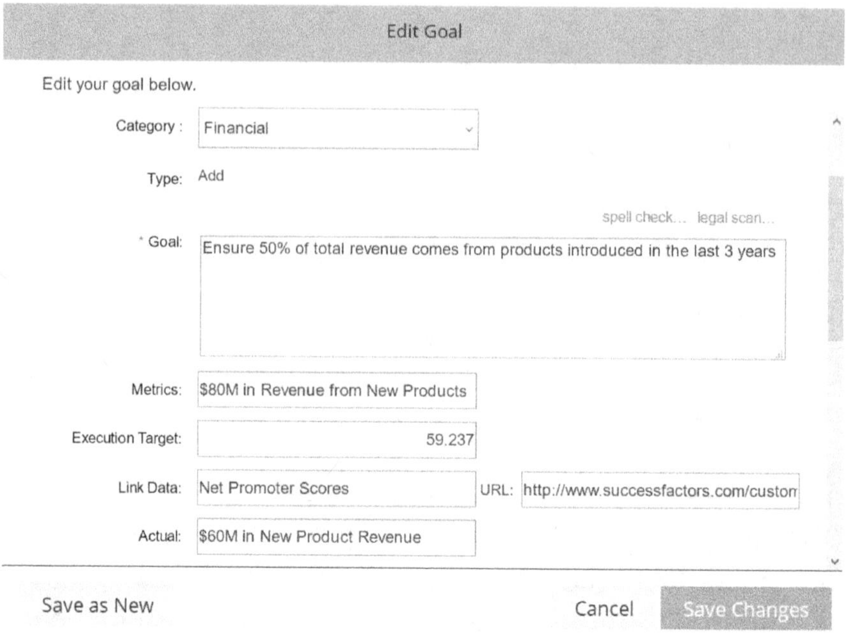

Edit Goal

Edit your goal below.

Category : Financial

Type: Add

spell check... legal scan...

* Goal: Ensure 50% of total revenue comes from products introduced in the last 3 years

Metrics: $80M in Revenue from New Products

Execution Target: 59.237

Link Data: Net Promoter Scores URL: http://www.successfactors.com/custom

Actual: $60M in New Product Revenue

Save as New   Cancel   Save Changes

Figure 1.4: Goal Plan in Edit Mode

Performance Management

The Performance Management (PM) module is where the actual performance review information is captured. It is tightly integrated with the GM module so there is no need to re-key any of the data that was entered into the goal plan. The template used to house performance management information is typically called a PM form template. This is also where your implementation consultant will be spending quite a bit of time, as the template must be designed and configured based on your organization's requirements and processes. One of the key features of the PM form template is that there is a workflow attached. This is also configurable, and it allows you to identify who will be included in the review and approval process for the employee performance review.

Figure 1.5 shows the team view for the Performance Management module displaying the current status of review forms that are in a manager's inbox.

My Team △ ❷	Feedback from Others ❷	Year-End Self Assessment	Year-End Manager Assessment	Calibration
 David Drew	You chose not to request feedback about David	Choose a rating 1.38	Choose a rating 1.2 Self Score Gap -0.18	1.0 Choose a rating
 Carol Clark		Choose a rating Choose a rating	4.0 Choose a rating	In Progress

Figure 1.5: Team View for Performance Management

There are different views that can be used when accessing performance review information. While Figure 1.5 shows the team view, you can also use the form view to see all performance review forms in a table format. These are conveniently categorized based on whether they are pending a manager's feedback, in another person's inbox, or are completed.

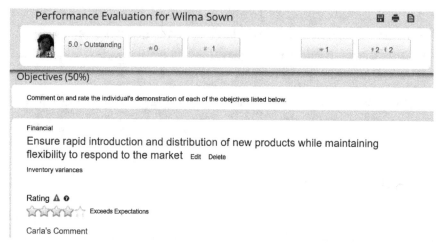

Figure 1.6: The Performance Review Form

Figure 1.6 shows an open performance review form. The different sections in a PM form are configurable, and your consultant will work with you on the design.

Functional Integration

Each of the modules in the HCM suite has built-in integration with other modules and the platform components. It is important to be aware of these cross modules integration so you can understand the dependencies between one module and another. At a macro level, dependencies can influence the order that the module is implemented. Understanding cross module dependencies is also critical when configuring the solution. This way the configuration design in one module does not limit the functionality that can be used in another.

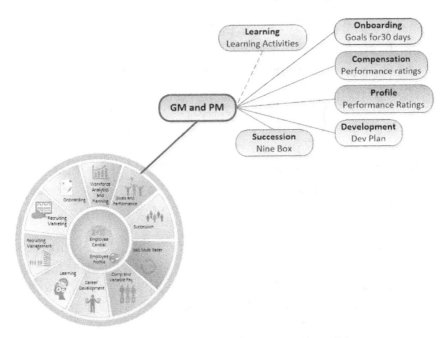

Figure 1.7: Functional Integration with
PM/GM and Other Modules

Figure 1.7 shows how the GM and PM modules are functionally integrated with several other modules. We will not go into this level of detail for the other modules. This is just to highlight that when designing your system, it is important to understand the interactions among the various modules from both an HR business process and a data perspective. For example, when one module is using a three-point scale and another a five-point scale, this will require some level of normalization and can create confusion for the users. Similarly, if the candidate profile is so different from your employee profile that the two of them cannot be synced, then this will impact your ability to seamlessly transition candidate data to employee data as part of the new hire conversion process.

Specific to the Goals and Performance Modules and the functional integration shown in Figure 1.7, the Compensation module is fed performance ratings such as overall goal rating, overall competency rating, and overall performance rating from the Performance Management module so this information can be used for compensation planning.

Overall ratings from the PM form can also be used in the Succession module when doing a talent search or to populate a 9-box (also referred to as a matrix grid, since some organizations will use a variant) when presenting performance versus potential or objectives versus competencies.

For those organizations that include development objectives during the performance cycle, this capability is also supported. The development plan can be embedded within the performance form similar to a goal plan.

All the overall ratings and the individual competency ratings from the PM form are stored in the employee profile and can be accessed by other modules or used to generate reports.

Finally, learning activities that are associated with an employee can also be visible in the PM form, but this will depend on whether the development plan is embedded in the PM form and if it is configured to display the learning activities. Similar functional integration capabilities exist across the various modules.

Succession Planning

The Succession Planning module helps you create, implement, and evaluate succession planning scenarios. It provides the features you need to plan potential personnel moves for key positions before they are needed. This give you the opportunity to mentor and develop employees so they can better succeed when moving into new positions.

There is a comprehensive set of features that can be used to rapidly identify potential successors, build successor pools, and manage the timing and availability of skilled successors. HR staff and managers can identify and define key positions along lines of business and organizational structures. Using this identification process, they can assign potential successors to these positions, either directly or from successor pools.

Figure 1.8 shows the succession organization chart. It is one of the primary tools that will be used to build out your succession planning program.

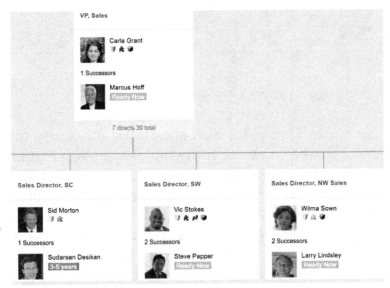

Figure 1.8: The Succession Organization Chart

There are many other features available within Succession Planning. For example, Figure 1.9 shows the succession lineage chart. This provides your organization with insight into the domino effect that could take place if a high-level succession plan is put into action. As nominees move into new positions, they in turn create vacancies. This chart helps managers understand the implications of executing a succession plan. The successors are arranged by readiness so the most likely scenario is shown straight across the top of the page.

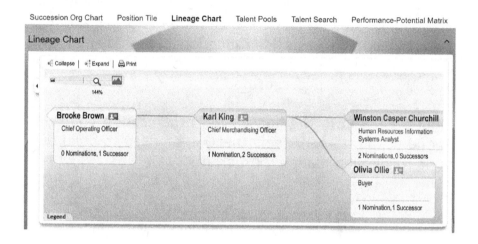

Figure 1.9: The Succession Lineage Chart

The top bar in Figure 1.9 also shows the other key tools that can be used to assist with your Succession Planning efforts. Most organizations will use at least one variation of the matrix grid (or 9-box). SAP SuccessFactors supports different variations of this report, such as performance versus potential, goals versus competencies, etc. Figure 1.10 shows an example of the top part of a matrix grid report. This can be configured during the implementation based on your internal requirements.

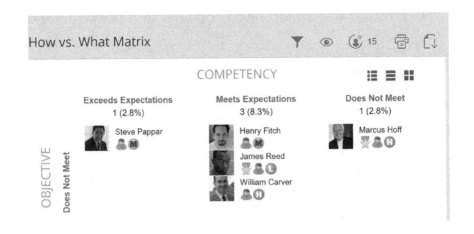

Figure 1.10: The 9-Box or Matrix Grid Report

360/Multi-Rater

The 360/Multi-Rater features are packaged within the Performance Management module. From a system architecture perspective, they use the same components. The use case however is very different, and that is why I have included this capability as a separate module.

This module provides a structured and confidential approach to requesting and receiving feedback on an employee's performance. While it is generally recommended that 360 be used for development, clients have used it for many other purposes. Here are some of the ways the 360 report can be used:

- Obtain structured and targeted feedback from colleagues

- Highlight development opportunities

- Aid career development

- Expose any underlying problems in a constructive way

- Facilitate development planning in the context of business

This feedback can be generated from a variety of respondents as highlighted in Figure 1.11. There is also the ability to use external raters, so you are not limited to feedback from within the organization.

Figure 1.11: Participants in a Multi-Rater Feedback

The two key components of the 360 module are the multi-rater form that is used as the input mechanism and the generated 360 report. Once the forms have been received and completed by the respondents, managers can view the 360 detail report. Figure 1.12 has an example of the 360 report.

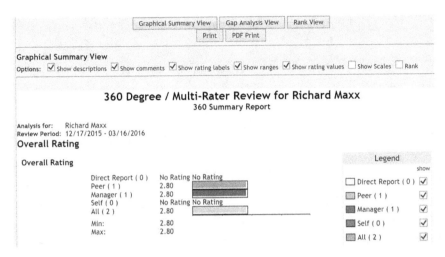

Figure 1.12: 360 Report

The top of the report has options to indicate how you would like to see the report. This is followed by the summary and then the details for the competency or area where a rating was done for the employee. The different variations of the report can provide some interesting perspective on the employee's capability. For example, if the blind spots view is selected, the report will display the information shown in Figure 1.13. As you can see, there are significant discrepancies between how the employee views her capabilities and how the raters view them.

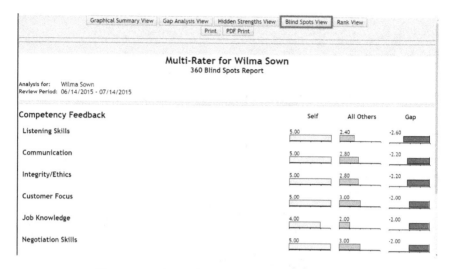

Figure 1.13: Multi-Rater Blind Spot View

This is only the first part of a comprehensive 360/Multi-Rater feedback process. How this information is ultimately used to improve the individual and the organization as a whole is a different conversation.

Compensation and Variable Pay

There are two modules available for compensation planning in the SAP SuccessFactors HCM suite. They are the Compensation and Variable Pay Modules. Both of these modules provide a very flexible set of tools that be used to support your compensation planning in an organization.

One of the first question clients will typically have is whether they need both of these modules. The Compensation module is used to complete merit planning and long-term incentives (equity), and Variable Pay is used to complete short-term incentives (bonus). To determine if the Variable Pay module is required, you will need to look more closely at how your bonus plans are calculated.

If your bonus plans account for time in position, then you will need to use the Variable Pay module to complete your bonus process. This will be a key decision in your project scope, so I have provided two examples below to help clarify. The first example is a simple bonus plan that does not account for time in position so the Variable Pay module is not needed.

A simple bonus plan would take an employee's annual salary and multiply it by a target payout percent to determine the bonus target. For example, $100,000 x 10% = $10,000. If the sample calculation shown here is how your organization determines bonus planning, the compensation module will be a good fit for your bonus and the complexity of variable pay may not be needed.

A more complex bonus would take into account factors during the year that account for time in position and requires proration to get to a blended payout total for the bonus. For example, let's add time in position to the previous example, so now we are looking at an employee who is promoted midcycle. For the first half of the cycle, the employee has the same above target. The second half of the cycle the employee is promoted, his or her base salary is increased by $10,000, and his or her target percent is increased from 10% to 20%. The calculation in this example would look like the following:

Bonus for first six months:	($100,000 x 10% = $10,000) x 50% = $5,000
Bonus for last six months:	($110,000 x 20% = $22,000) x 50% = $11,000
New Bonus Target:	$16,000

If the above calculation looks familiar, or your organization accounts for time in position for business results, you will most likely need to use the Variable Pay module. As you go through the scoping process, your SAP SuccessFactors representative can also provide guidance to help you determine if Variable Pay will be needed.

Once you have scoped out if both Compensation and Variable Pay are needed or if you only need the Compensation module, the next step is to focus on the implementation. I have covered the implementation process in more detail in later chapters, but I wanted to highlight some of the unique change management, data requirements, and cross-module dependency considerations when implementing the compensation modules.

Due to the sensitive nature of compensation, any changes to the current tools and process can give your compensation team a high level of anxiety. I recommend spending time up-front preparing them for these eventual changes to their tools to help with the transition to the SAP SuccessFactors system. This will help them be more comfortable that the new system has all the capabilities to help them plan, audit, report, and administer their compensation process.

The data requirements for compensation are also unique. Compensation requires clean, up-to-date data for all fields that feed to and from the compensation form. Thinking about how your organization currently compiles and audits employee data may adjust how you approach your SuccessFactors implementation. First consider what data sources will provide employee data: Do you have multiple HRIS systems? What about equity—is that data with a separate vendor, like Fidelity?

For bonuses, do you have business results that come from fi nance? If you have purchased and plan to use SuccessFactors Employee Central tool for HRIS and want to integrate the compensation forms with EC, then all the employee indicative data will need to be sourced from EC. Therefore, you will need to refi ne the implementation approach to ensure that EC is live before Compensation is implemented.

For cross-module dependencies, I had provided an example earlier in the Performance Management module section, but I want to highlight here again because it is critical to compensation. Besides EC, goals and performance ratings from these SAP SuccessFactors modules will feed performance scores into compensation and variable pay, so consider rating scales and how performance scores are calculated.

To wrap up the overview of these two modules, let's look at the compensation module in a bit more detail as this is more frequently implemented than the Variable Pay module. Figure 1.14 shows the compensation plan for a manager. Details for each direct report and summary information are presented. At the top of the form, the user can indicate whether they would like to see budgets, metrics, or route map information. In Figure 1.14, the metrics option is selected. This view also supports drill-down capabilities for additional details on the employees.

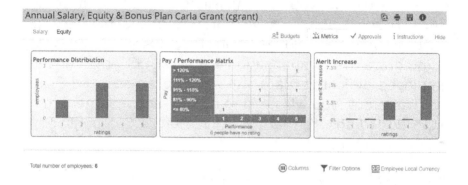

Figure 1.14: Compensation Form Metrics Section

Once you select a specific individual, you can click on their rating link to access their performance review form. Alternatively, clicking on their name will display more specific information for the employee's compensation profile. This is represented in Figure 1.15. Many of these fields and the approval process are configurable. As part of the implementation process you will work with your consultant to determine which fields to include and how your compensation process will be represented in the system.

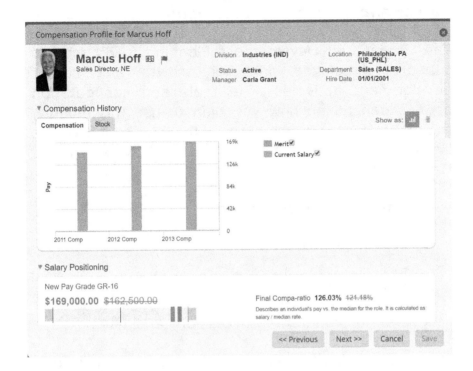

Figure 1.15: Compensation Profile

As you start planning for the Compensation implementation, do factor in some additional time to review your compensation planning process and philosophy. You could use this time to review your compensation philosophy with your teams. Overly complex calculations on compensation fields can confuse planners and employees. Think what your organization is trying to drive through incentivizing behavior and whether or not employees understand what you are trying to get them to do. Keeping the planning process simple makes the process easier for you to complete strategic compensation decisions, like reviewing budgetary allocations across the organization.

Career and Development Planning

The Career and Development Planning (CDP) module supports employee personal growth and career development activities. The tools and configuration options depend on how you plan to use this module, the maturity of your competency model, and if it will be integrated with a Learning Management System (LMS). Here are the primary use cases:

- Employee personal growth

- Career planning

- Correctional measures

- Developing internal succession candidates

This module relies heavily on the competency framework. Even the linkage of development goals to a specific learning activity in the LMS is based on competencies. My general recommendation is to initially take a basic approach when using this module. Once your competency model is completed, you can enable and start using the advanced features. Figure 1.16 shows the main page of the Development module. Additional features such as the career path are displayed in the top menu bar.

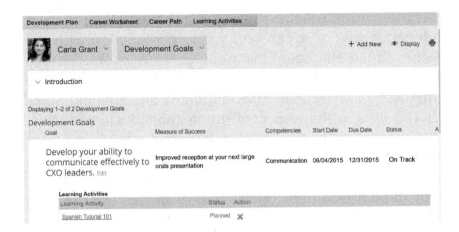

Figure 1.16: Main Page of the Development Module

The development goal can be modified in the detail view. Figure 1.17 shows the edit view of the development plan.

Edit Development Goal

Edit Development Goal

* Goal:
Develop your ability to communicate effectively to CXO leaders.

spell check... legal scan...

Measure of Success:
Improved reception at your next large orals presentation

Start Date: 06/04/2015

Due Date: 12/31/2015

Status: On Track

* Competencies:
☐ Acting as a Champion for Change
☑ Communication
☐ Customer Focus

Cancel   Save & Close

Figure 1.17: Edit View of the Development Plan

For the initial rollout, the CDP module can be used independently, or it can be used in conjunction with the Performance Management module. Many clients embed a development plan in the performance review form so employees can set their development and training goals for the year while also confirming their business goals. Often, when rolling out a development plan, it is easier for a consultant to make a duplicate copy of the goal plan or to add another category in the existing goal plan for development goals. It is a better approach to always use a development plan from the CDP module for development activities, even if it is just a basic version. This approach will be much more scalable, allowing you to add features to the basic development plan instead of having to replace the modified goal plan that was initially used for development.

In later phases, once the competency model has been defined, the career worksheet and career pathing can be rolled out. Figure 1.18 shows a sample view of the career worksheet.

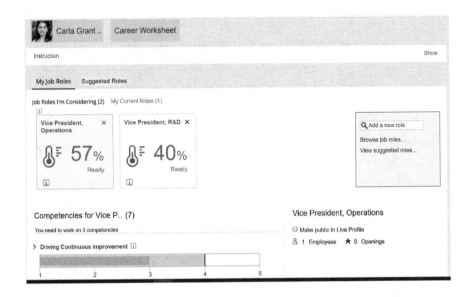

Figure 1.18: The Career Worksheet

Learning Management System (LMS)

The Learning module is also known as the SuccessFactors Learning Management System (LMS). This module provides a scalable and flexible learning platform that can be used to provide classroom, web-based, and mobile device training. It has an intuitive user interface as depicted in Figure 1.19 that makes learning activities easy to access and track for internal and external audiences.

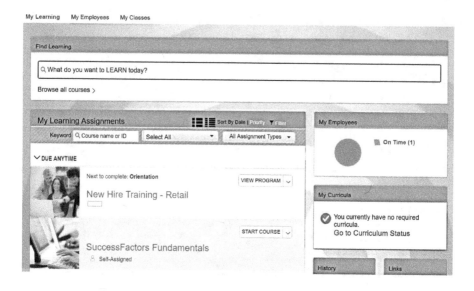

Figure 1.19: Landing Page for the Learning Module.

The LMS has many features and functionalities, making it a good fit for basic and advanced learning management needs. Some of those features include:

- delivering online content to learners

- learner self-registration for instructor-led courses

- supervisor assignment of courses to learners

- automatic assignment of courses based on user attributes

- mobile access to learning

- integration with a virtual learning service (VLS), such as WebEx or Adobe Connect

- instructor access to rosters and marking attendance that does not require administrator rights

- tracking and reporting learning completions for online and instructor-led courses

- eCommerce functionality that can allow for credit card payment for courses

- automate and track compliance/certification training

- design and deliver blended learning paths

While this is not an exhaustive list, it does highlight the different options and configuration decisions that need to be made when implementing the LMS. Some companies will use the LMS only for training their employees, but many clients will also use it to train their vendors and customers, as external audiences can also have access to the learning content. A few clients will use the LMS to train their employees on other SAP SuccessFactors modules to support the implementation. This, of course, will depend on whether the LMS is implemented early in the project before some of the other modules in the HCM suite.

Some configuration decisions will be dependent on how the LMS will be used. Is it to support compliance, skill building, or just-in-time learning for employees and clients or all of them? This will have an impact on the type of learning activities that will be included in the LMS and the user data needed to support the implementation. For example, if vendors will participate in training, then the vendor's employees will need user accounts to have access to the LMS.

The SAP SuccessFactor leaning module supports seamless integration with the other modules in the suite and also with third-party content management system. Learning activities that are associated with a development goal in the CDP module can be linked to courses in the LMS. The content for these courses can exist in the SAP SuccessFactors iContent server or in a third party system. Figure 1.20 below shows the setup screen where you can indicate the location of the content. This screen also shows that all the popular content standards are supported such as AICC and SCORM.

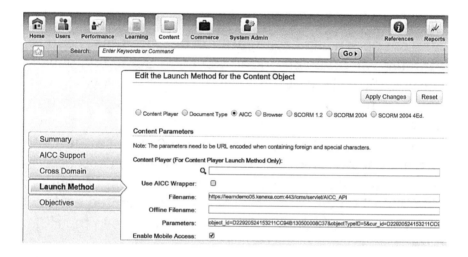

Figure 1.20: Managing Learning Content

Jam

The Jam module is used for social collaboration and learning. There are countless uses for this module as it has some very flexible capabilities. You can use it for social networking where employees and clients can share information in multiple formats and relationships. Information can be shared to a single person, a project group, teams, or departments.

Similar to the LMS, it can also be used to support the implementation by serving as a centralized repository for all the project documentation. The nice thing about using it this way is that the project team is actually using the tool to support the implementation while becoming more familiar with it. For this reason, Jam is normally implemented first in many implementations. Figure 1.21 has a screenshot of the main page.

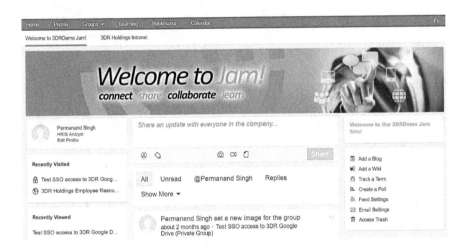

Figure 1.21: SAP Jam for Social Learning and Collaboration

One of the really nice features is the ability to create content-specific Jam groups to encourage interaction and knowledge sharing across a team or group. The UI can be modified to make it very interactive and user-friendly. Figure 1.22 has a screenshot of a manager portal that was created by the 3DR team to support managers. As you can see it has user-friendly images that can be used to drill down to additional level of details.

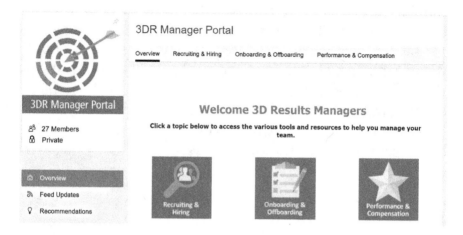

Figure 1.22: Manager Portal

The traditional file folder structure is also supported, so if you want to look as specific folders or search for files this can be done. Figure 1.23 shows how you can look at Jam content using the folder structure. As mentioned before this is a very versatile tool and can be used to train your users, store and share information on your SAP SuccessFactors project implementation, or use as a manager portal to mention a few use cases.

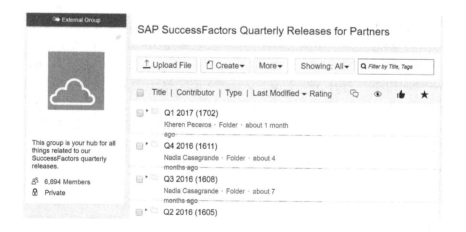

Figure 1.23: Jam Folder Structure

Recruiting Marketing

The Recruiting Talent Acquisition modules, provided by SAP SuccessFactors, focuses on sourcing, engaging with, and hiring talent for organizations around the globe. It's above and beyond just your average applicant tracking system (ATS) and provides robust capabilities that enable an optimized candidate experience while providing innovative tools for organizations to attract, engage, and hire talent.

Within the SAP SuccessFactors HCM suite, there are two separate yet fully integrated modules. These are the Recruiting Marketing and Recruiting Management modules. I will cover the capabilities of the Recruiting Marketing module first in this section and then the Recruiting Management module in the next section, but first let us take a look how organizations use SAP SuccessFactors to get the word out about job openings so that candidates can find and apply to these positions.

Recruiting Postings

SAP SuccessFactors provides the ability to post jobs to more than three thousand sources in over eighty countries, and when candidates apply from these sources it's all electronically tracked.

Two common approaches to distributing job postings is to either set up regular job feeds directly to electronic sources (considered automated sourcing) or to leverage SuccessFactors Marketing Central where hiring representatives can pick and post to specific electronic sources on an ad hoc basis. It's important to use the system-delivered capabilities in order to ensure proper candidate source tracking takes place so the organization can understand effective sources.

An organization's job distribution needs may come at an additional cost, so your implementation consultant will guide you through these decisions during the implementation. In addition, SAP SuccessFactors provides a media management team for organizations seeking extra support in expanding their reach and taking their advertising efforts to new heights.

The Recruiting Marketing (RMK) module provides comprehensive features like a fully hosted careers site that is powered by jobs posted in Recruiting Management to attract and engage external candidates. RMK is often considered the crown jewel of the Recruiting Talent Solution as most competitive solutions have no such offering to compete against it. This module dynamically markets and presents jobs to external candidates, automates engagement with those candidates, and provides visibility into data that answers some of the most common questions faced when acquiring talent such as the effectiveness of each candidate source.

The most visible feature of RMK is an organization's career site. The site can be built using either SuccessFactors web designers and developers (which is at a higher implementation cost and will require the organization to work with SAP SuccessFactors for post go-live changes to the site) or an innovative tool called Career Site Builder (which is a lesser implementation cost and enables the organization to update the site themselves post go-live). Which way your site is built has different benefits, and the decision is typically made during the sale of implementation services but could be changed early in the implementation effort if so desired by the client.

Figure 1.24 Career Site Built Using Career Site Builder

Regardless of which way your site is built, the following key features of RMK remain intact:

Search Engine Optimized (SEO) Career Site

SAP SuccessFactors holds a patent on a technology that builds your site's relevancy on the major search engines (i.e., Google, Bing, etc.). This is accomplished by the platform spinning off additional SEO pages based on your jobs relevant keywords and locations with the goal that every single job sits on a searchable web page for candidates to easily access. The platform also indexes the searches candidates are making on your RMK site and stores them on a searchable web page.

This technology's goal is to drive applicant flow and get your jobs in front of candidates who conduct searches on these sites, particularly those who may never visit your career site directly.

Responsively-Designed Career Site

SAP SuccessFactors career sites are built to automatically present the site in the most optimal way based on the size of the candidate's browser and device.

This way the candidate's experience from browsing the site through to applying for the job in Recruiting Management can be one seamless and user-friendly experience.

Talent Community Marketing

One of RMK's most valuable features is its ability to capture candidate interest in an organization, or a position, and to enable an organization to remarket to those candidates either automatically or on demand by a hiring representative.

Every time a candidate visits the career site they will be prompted to join a talent community, and every time a candidate applies they will automatically join the talent community. Upon joining the talent community automatic job agents are set up in the system that records the candidate's interest.

This database grows exponentially post go-live and commonly becomes one of an organization's top sources—and it's freely included in the module and runs automatically, without administrator support.

RMK Dashboard

This feature enables hiring representatives to proactively pursue talent and to have quick data at their fingertips about their site's performance and any marketing efforts results.

From this tool, hiring representatives can search talent community members, generate campaigns, e-mail candidates, have quick access to many useful statistics, and generate URLs/QR Codes/Social Links for jobs that can be posted on other third-party sites.

RMK Advanced Analytics

This feature marries the site's performance and marketing efforts data, the information available in the RMK Dashboard, with information from the Recruiting Management module.

This provides visibility to data from the job that was posted through to when the candidate was selected and hired—allowing hiring representatives to identify which new hires came from which sources. This powerful information is aligned to the cost of marketing jobs thus providing organizations the opportunity to drive effective marketing decisions using these metrics.

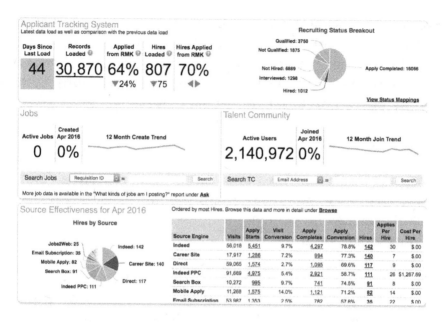

Figure 1.25: RMK Advanced Analytics Homepage

Recruiting Marketing Implementation Considerations

The implementation of Recruiting Marketing is remarkably simpler than Recruiting Management and is often done in parallel when implementing both modules. Instead of needing to make decisions that impact business processes like in Recruiting Management, when implementing Recruiting Marketing the majority of input an organization has is related to decisions on their career site's design—in short, its look and feel and the words displayed on the pages.

Typical the RMK career site includes a home page, content pages where the organization can provide details about itself, strategy pages broken down by organizational function that provides quick access to those related job postings (e.g., IT, Human Resources, etc.), a location map (which allows quick access to jobs by navigating a regional/national/global map), and the job posting page itself. All of these pages require branding—from company logo, colorization, images, and photos.

It's important to remember that the number one goal of your site is for the candidate to easily search, locate, and apply for positions with very few clicks. Candidates that are directed to your RMK career site from an electronic source rarely take the time to visit other pages. With this in mind, when building your site exercise restraint with how complex the site is to navigate and with how many pages exist on the site. This will avoid the pitfall of a candidate getting lost on the site and not applying for positions in your company.

Prior to an implementation of Recruiting Marketing it's also helpful for an organization to have a handle on their employment brand so that the career site, the first impression to candidates your organization makes, clearly articulates and reflects your brand. This is important because the organization is completely in the driver's seat when it comes to the careers site's design, and it's often useful for implementation teams to engage appropriate individuals within their organization to support this effort (e.g., marketing, internal communications, current site's webmaster, etc.).

Last but not least, it's important for organizations to vet design ideas and mock-ups (provided in the implementation effort) with organizational leadership to ensure all are in the know. The careers site is an organization's first impression with candidates, so be sure that all opinions are gathered internally while listening to the guidance of the implementation consultant.

Recruiting Management

The Recruiting Management (RCM) module has a diverse set of capabilities to streamline and expedite requisition, application, candidate, interview, offer, and eventual hiring processes. It is equally balanced with tools to support efficiency for hiring representatives while also providing the internal candidate experience when an existing user is logged into SAP SuccessFactors.

The main features of the module include:

Requisition Management

Requisitions (reqs) can be created within SAP SuccessFactors in a variety of ways. The most common is to create them based on job profiles that can prepopulate the job description and job/position code details to streamline the effort of completing the req.

Requisitions can then be routed for approval to individual participants identified on the req or on an ad hoc basis to ensure all involved are informed of the new opening. Upon final approval, requisitions are available to post, which pushes the position to RMK (if implemented). Fields on requisition are typically limited to just those fields necessary to inform the hiring representative, to be used for the job posting, to be used to power reporting, and to be used to feed to downstream systems.

Requisitions are then accessible to any requisition participants identified on the req and are displayed on the Recruiting dashboard that is accessible when a user is logged into SAP SuccessFactors. The Recruiting dashboard has various display options for each user to select based on their preference. The screenshot below in figure 1.26 shows an example of a Recruiting dashboard.

Figure 1.26: Recruiting Dashboard

Internal Careers Site

Users with access to log in to SAP SuccessFactors and permission to the internal careers site are able to search for jobs using keywords and other organization-defined criteria.

In addition, users are able to set-up job alerts, save jobs, save job applications, view statuses of previously submitted applications, view their candidate profiles (which is intrinsically linked to their employee profiles so that if one is changed it updates the other), and submit external referrals for jobs they may be qualified for (which allows and organization to have an automated and reportable employee referral program). The figure below shows the different features and menus that can be accessed from the internal careers site.

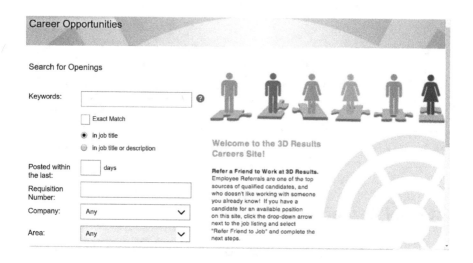

Figure 1.27: Internal Careers Site

Application Management

When external candidates apply using Recruiting Marketing or when internal candidates/employees apply using the internal career site, both go through a fairly similar experience. After reviewing the job posting, the candidate is first presented with their candidate profile, which allows for dynamic capture of background information (e.g., employment, education, etc.) and then is moved to a job-specific information section for capture of information for the type of job, country, and job posted. This is also where candidates are prompted to complete any prescreening question(s) identified on the req—these questions can be used to knock out applicants, to rank applicants, or to simply gather information.

External candidates are automatically set up with a job agent when they apply via Recruiting Marketing, and internal candidates have the same capability using the internal careers site.

Both external and internal candidates are allowed the opportunity to save jobs, save their work midapplication, view their candidate profiles at any time to make updates if so desired, and to view the status(es) of previous applications.

Candidate Management

Applications by external, internal, referral, and agency candidates are made available on the candidate tab of each open req. Hiring representatives are able to click on the candidate's name to view their candidate workbench, drag/drop or click the application to manage their status, send ad hoc e-mail correspondence, and take action on multiple applications at the same time if so desired.

In addition to managing candidate statuses, the module also provides security controls on what a participant on a requisition is able to see or take action on based on the type of participant they are (e.g., recruiter, hiring manager, etc.). Candidate statuses also provide robust capabilities in driving automated or ad hoc e-mail communications to candidates and requisition participants, ensuring all are kept informed on the latest status of a candidate's application. Figure 1.28 shows a screenshot of the candidate dashboard where a recruiter can get details on the candidate and their application.

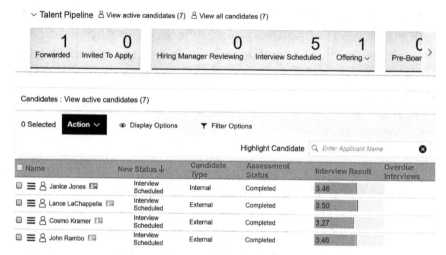

Figure 1.28: Candidate Dashboard

Offer Management

SAP SuccessFactors provides many capabilities when it comes to creating and delivering offers to candidates. Not only can offer details be captured in the system and routed to internal users for approval, but offer letters can be generated and delivered to candidates in numerous ways.

Some of the legacy delivery methods include printing out the offer letter (while marking it as printed), providing the offer verbally (while marking it as a verbal offer), sending the offer as the body of an e-mail, and sending the offer as a PDF attached to an e-mail.

In addition to those, there is the capability of delivering the offer letter online while providing the candidate the opportunity to select a button indicating their offer response, and/or there is the capability of delivering the offer letter via a standard integration with DocuSign, which provides the candidate and your organization the opportunity to obtain a true electronic signature on the offer letter (DocuSign contract is required to leverage this feature).

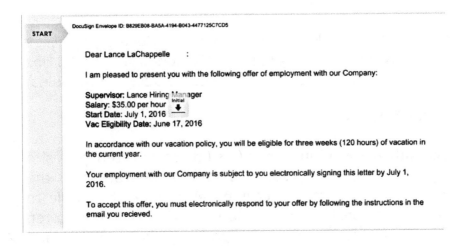

Figure 1.29: Offer Acceptance with eSignature

Hiring Candidates

After candidates accept their offers, SAP SuccessFactors provides multiple ways to proceed in processing a new hire (or even an existing employee's job change if so desired by the organization). This can eliminate the duplicative task of keying a new hire into an HCM/payroll system.

These ways include passing any data from the Recruiting Talent Solution into:

- SAP HCM (leveraging SAP add-on connectors powered by SAP Process Integration (PI) or SAP Hana Cloud Integration (HCI) platforms)

- SAP SuccessFactors Employee Central HCM Core HR Solution (leveraging internal SAP SuccessFactors data mapping capabilities)

- SAP SuccessFactors Onboarding Talent Solution (leveraging internal SAP SuccessFactors data mapping capabilities)

- other HCM/payroll systems (leveraging .csv exports from SAP SuccessFactors using the Ad Hoc Reporting and/or Online Report Designer modules)

Integrations

As with many hiring solutions, there is often a desire by clients to build real-time integrations between the SAP SuccessFactors Recruiting Management module to other third-party vendor solutions. These vendors often support the administration of pre-employment behavioral assessments, Work Opportunity Tax Credit (WOTC) screening (in the United States), virtual interviewing vendors (e.g. HireVue), pre-employment screening (background checks, drug tests, reference checks, etc.) and more.

Real-time integrations are built using SAP SuccessFactors OData APIs (Application Program Interface), which allow third-party middleware tools (Dell Boomi, Hana Cloud Integration, Cast Iron (IBM), custom solutions, etc.) to query, select, and export data from the SAP SuccessFactors database. In return they also allow the input of data into the database from third-party vendor solutions (e.g., background check results, behavioral assessment results, etc.).

SAP SuccessFactors provides standard integrations to two pre-employment behavior assessment vendors that run on the Dell Boomi integration platform, which the implementation partner must implement. These integrations are then supported by SAP SuccessFactors post go-live.

Integrations to all other vendors is the implementing organization's responsibility to have configured with their implementation partner (or internally). Integrations are typically built using Dell Boomi technology; however, other tools that can interact with APIs can be used as well. These integrations must be supported by the organization's implementation partner (or the client) post go-live.

Figure 1.30: HireVue (Virtual Interviewing Vendor) Status Displayed on Candidate Workbench

Automation

For companies with a high volume of candidates, there exists the opportunity to automate actions typically completed by hiring representatives. An example of this automation includes moving a candidate to a different status when his or her background check result is returned, which can then send off an automated notification to a hiring representative.

Third-party integration tools can be used to monitor data (by connecting to the OData APIs) within the database and be programmed to follow a set of rules to take actions within the system when specific criteria is met (e.g., if result equals pass, move candidate to *screening cleared* status).

Recruiting Management Implementation Considerations

When implementing Recruiting Management it's important for a project team to include your typical project manager, project sponsor (e.g., head of talent acquisition/recruitment), project business lead (e.g., talent acquisition/recruitment operations manager), IT representatives (if any security protocols need to be addressed and/or any support integrations to internal organizational systems are required), and a cross section of a few hiring representatives that would be using the system regularly. It's important to engage all these participants in the design, development, and testing phases of the project to ensure success. It's also useful to then leverage this team as change agents as the go-live approaches.

The complexity of implementing Recruiting Management can vary greatly from one organization to another because it enables an entire process that has multiple participants, not just a transaction. It is also used to complete one of the most important tasks Human Resources departments can provide—hiring quality talent. This is why proper project resourcing will be required to support the effort.

Two main drivers of implementation complexity include:

- *Distinct Hiring Processes*

  The majority of organizations typically have multiple hiring processes. Examples include: the corporate office has a team of recruiters who support all hiring activities in that building, but in the field offices or retail stores the hiring managers may be tasked with doing their own hiring; one portion of the organization uses one HCM/payroll solution and the other uses a different system; one part of an organization uses a heavy contractor/agency process and needs a different candidate status workflow.

The quantity of hiring processes can impact the quantity of components that make up the module, thus increasing the configuration, unit testing, and user acceptance testing required. These components include requisition forms, requisition approval workflows, candidate application forms, candidate status flows, offer approval templates, and so forth.

- *Integrations*

  Beyond just the quantity of integrations that may be required, the method of building them impacts complexity based on who will be responsible for building the integration and the length of time the implementation takes. For example, if the implementation partner is building the integrations and supporting it post go-live, then this would require little involvement by the organization's IT department.

  Oftentimes, organizations will work with the implementation partner to build the integration, and the organization's IT department would build any interfaces of that integration to internal systems (e.g., HCM/payroll solutions, etc.).

  It's important that prior to finalizing the sale of services the implementation partner will provide that any desired integrations are raised to the implementation partner and an organization's IT department.

It is important, as with any implementation of a SuccessFactors module, to thoroughly test all requisition templates, offer approval templates, application forms, and the candidate profile. This is important because post go-live organizations are unable to change the content of these items (such as field labels, fields, field attributes (e.g., is it required or not, drop-down versus text, etc.)) without a professional services organization's support. There are many things an organization's system administrator can modify post go-live (e.g., e-mail communications and when/who they are sent to, drop-down field options, deleting/reassigning requisitions, and much more).

Although it may seem overwhelming, a qualified implementation partner will help organizations in making the right decisions for the module—striking a delicate balance between complexity of the solution and ease of use for hiring representatives and candidates.

Onboarding

The Onboarding (ONB) module can be used as a stand-alone module, but when used with the Recruiting Talent Solution it will drive additional efficiencies across the organization that may have hard costs associated to them. The Onboarding module provides the workflows and components necessary to effectively and efficiently onboard new hires, cross-board existing employees into new jobs/locations/regions/etc., and off-board existing employees who leave the company.

Workflows

Now it may seem strange that all three of these actions are included in this one module, but it's because of the module's advanced workflow capabilities, document population/retention capabilities, and its inherent integration to other SAP SuccessFactors modules (i.e., Recruiting Management and Employee Central).

Onboarding Workflow

This is the most prevalent workflow leveraged in the Onboarding module and is used to provide a consistent and efficient method of converting new hires into engaged, empowered, and productive employees.

In this workflow, information about the new hire can either flow over from the Recruiting Management module, be entered manually, or flow in from another applicant tracking system (when implementing as a stand-alone module without Recruiting Management).

Standard workflow steps include a post-hire verification step that allows a hiring representative to confirm/update information about the new hire, a new employee step that provides the new hire the opportunity to confirm/update/ provide information about themselves while also viewing/ signing any required documents, and an orientation step that is assigned to the hiring representative and used to confirm the employee's arrival on their first working day.

Cross-Boarding Workflow

As the next most prevalent workflow in the Onboarding module, this is used to ensure a consistent and efficient method of capturing data from an employee transferring from one position/job/location to the next while also delivering the employee information that is relevant based on their change.

In this workflow, information about the new hire can either flow over from the Recruiting Management module, flow over from the Employee Central module, be entered manually, or flow in from another system (when implementing as a stand-alone module without Recruiting Management or Employee Central).

Standard workflow steps include a post-hire verification step that allows a hiring representative to confirm/update information about the employee's new position, and a new employee step that provides the new hire the opportunity to confirm/update/provide information about themselves while also viewing/signing any required documents based on their new position.

Off-Boarding Workflow

The final workflow this module enables is used to ensure a consistent method of off-boarding an existing employee whose employment with the organization is being terminated.

In this workflow, information about the new hire can either flow over from the Employee Central module, be entered manually, or flow in from another system (when implementing as a stand-alone module without Employee Central).

Standard workflow steps include an initiation step completed by an assigned resource (e.g., HR generalist, hiring manager) to confirm/update about the employee, an employee step that provides the employee the opportunity to be informed of organizational protocols that must be followed (e.g., return of equipment, links to exit surveys, etc.) and also to view/sign any required documents, and an exit interview step completed by whomever is seeing the employee on his or her final working day.

Components

With each of these workflows the implementation partner and organization has a variety of components that can be used.

Panels

The module leverages pop-up windows (referred to as panels) to present information and capture information from users. These panels can be prepopulated with information, present content, present hyperlinks to documents/other websites/etc., and even link to a video.

Panels themselves and their fields and content displayed can be controlled using advanced conditional logic that looks at any data in the module (that was either passed from another module or entered directly) to determine what is displayed to the end user. Examples include only showing a specific policy form to new hires starting within a specific location, only asking a specific question based on the company the new hire is being employed by, etc. Figure 1.31 shows a sample post-hire verification step panel.

Figure 1.31: Post-Hire Verification Step Panel

Forms

The Onboarding module provides fairly advanced form/documentation management that can be controlled, just like panels, based on advanced conditional logic.

Forms can be populated with or without data and then can be administered to internal or external users in a variety of ways, including: displaying forms on a panel while requiring a user to view it, attaching forms to e-mail communications, presenting forms to be signed using either SuccessFactors' legacy Click to Sign feature or an innovative new Click to Electronically Sign feature using DocuSign (DocuSign contract required to leverage this feature).

All forms are then retained in the module's document repository for an unlimited time, allowing users with the appropriate permissions to access/print these forms when/if necessary. If needed, forms from the document repository can be exported so that they could be imported into an organizations existing electronic personnel file management system.

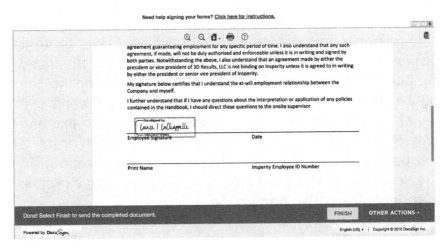

Figure 1.32: Form Signing Process

Portal

This tool provides the user—a new hire (in the case of Onboarding), an existing employee (in the case of Cross-Boarding), or a terminating employee (in the case of Off-Boarding)—with a website where they can view information relevant to the workflow. This website, like a recruiting marketing career site, can be configured based on an organization's brand and can have numerous pages to cover the necessary content that is relevant for the user.

This portal is a site that helps personalize the employment experience by providing important information before their first day. The portal helps guide the pending hire through all the resources available to him or her as he or she prepare for the transition to the organization (e.g., benefits information, policies, procedures). In addition, the organization can place videos on the site, provide background on the company, and/or introduce the pending hire to the organization's core values, mission statement, and corporate objectives. If the content exists, it typically can be published on the site for the pending hire's benefit.

Like other components in Onboarding, the content displayed on the website can be controlled by advanced conditional logic—it's all about displaying content to the right users at the right time (e.g., a new hire can't see benefit costs until after his or her start date, an existing employee change to a new location provides him or her a webpage that provides directions to the new location). In the screenshot shown in Figure 1.33, I have an example of a new hire portal homepage.

Figure 1.33: Example of New Hire Portal Homepage

Other Onboarding Workflow Components

In addition to the above components that apply to all workflows and require the most configuration effort, your implementation consultant will walk you through other key component options that an organization may want to use. These include:

- providing new hires access to SAP SuccessFactors in advance of their first day;

- getting new hires instructions and access to the SAP SuccessFactors mobile app before their first day to prepare them for joining;

- leveraging the Onboarding Tour, which acquaints new hires with SAP SuccessFactors modules when they first log in after obtaining access; and

- activating the hiring manager checklist, which allows them to complete certain actions in SAP SuccessFactors or on its mobile app for a new hire (i.e., assign a buddy to the new hire, recommend other employees to the new hire, set up meetings, set up his or her new hire goals (requires a new hire goal plan be configured in the Goal Management module, provision complete a checklist, etc.)

Integrations

After the new employee step is completed in either the On-Boarding or Cross-Boarding workflows, information about the new hire can be mapped over to the Employee Central module (if implemented) or exported from the system for load to another HCM/payroll systems (in a .csv or .xml format). This helps organizations avoid the duplicate entry of a new hire's information and the potential data integrity issues that could arise.

Onboarding Implementation Considerations

The complexity of implementing Onboarding varies greatly based on the quantity of the aforementioned workflows and components (e.g., panels, forms, etc.) and how robust the portal will be (quantity of pages/ design and thus content). The number one priority for an organization to prepare in advance of implementing the Onboarding module is to get their arms around the various forms that are used in these steps of the process, to evaluate those forms to see if there is any duplication of them across the organization that could be mitigated through consolidation, and to design new forms that appropriately represent the organization's brand.

If implementing Recruiting Management, the implementation of the Onboarding module and workflow is typically done in parallel because the audience impacted by the installation of the Recruiting Talent Solution and the Onboarding module are typically the same. This then reduces the complexity of training/change management activities while also providing for huge efficiencies in organization without electronic paperwork in place prior to this implementation. Use of the cross-boarding, and most definitely the off-boarding, workflows is done typically after initial go-live.

SAP SuccessFactors maintains certain standard compliance processes and forms (availability varies by country) for organizations to leverage. Some examples include the following, but ask your implementation consultant about when these standard processes and forms are right to use for your organization.

- In the United States, Form I-9/e-Verify and federal and state tax forms

- In Australia, federal tax forms

- In Canada, federal and provincial tax forms

During the implementation of Onboarding the project team should include a project manager, project sponsor (e.g., head of HR operations/HRIS, etc.), project business lead (e.g., HR operations manager), IT representatives (particularly if integrating to organizational systems is desired), and a cross section of representatives involved in the paperwork administration and HCM/payroll system transactions. It's also important to include individuals from various areas based on what is being implemented in the module. For example: the tax department should be included if planning on implementing tax forms, the payroll department should be included if planning on having payment methods (e.g., direct deposit, PayCard enrollment, etc.), the internal communications/learning team should be included for support with portal content/imagery/etc.

Post go-live, an organization's system administrator is able to make many changes to this module, including but not limited to adding/editing/removing panels, adding/editing/removing forms, updating the portal, adding/editing/removing e-mail communications, etc.

Workforce Analytics and Planning

Today, more than ever, technology has provided companies with massive amounts of data about their employees. This data can be put to use not only to manage the administrative tasks related to managing and paying employees, but also to gather insights and knowledge about a company's most important asset: their people. Being able to make effective business decisions about this asset is critical to the success of *any* company. And the path is littered with the remains of companies who have mismanaged this critical asset. Thus the criticality for organizations to understand and effectively manage their workforce. Succinctly put, Workforce Analytics is using that employee data to make better, more insightful decisions about the workforce.

Workforce Analytics improves workforce decision-making by helping your company find answers to key questions about workforce challenges and how to solve them. This module comes with over ninety key performance indicators (KPIs). With this information, organizations can identify trends at an early stage and make well-informed decisions to manage their human capital more effectively, predict human-capital investment demands, and track workforce costs.

Clients can typically use the Workforce Analytics component to accomplish the following:

- identify key attributes of your talent pool

- measure recruiting processes

- monitor performance management processes

- measure employee compensation programs

- research and solve emerging workforce issues

- measure and analyze learning programs

- analyze succession programs

- measure and analyze typical core HR operational processes, such as payroll, employee administration, time management, and benefits

There are metrics and dashboards that can be developed with this module. Figure 1.34 shows a sample screenshot of this module.

Figure 1.34: Workforce Planning and Analytics

SuccessFactors' Workforce Planning module is used for reporting and analysis to gain insight into your workforce needs. This module helps you determine the type and amount of talent you will need to reach organizational goals over the next few years.

With the Workforce Planning module, you can:

- Understand current workforce trends and plan future needs by using workforce demographic data.

- Use more than 200 predefined reports to analyze headcount development, turnover rates, and workforce composition.

- Link the results of this analysis directly into headcount planning, budgeting, and key talent processes, such as recruiting and learning.

- Get access to a broad range of workforce-related data to support accurate planning, facilitate simulated planning scenarios, and enable continuous monitoring of actual performance relative to plan.

Employee Central

SuccessFactors Employee Central is a flexible and highly configurable HR management solution that sits in the cloud and offers intuitive technology for organizations to effortlessly manage their employee information. There is more to the system than simply automation of processes, and as such, it has developed into a solution that can be used worldwide, enterprise-wide, and through every level of user in an organization.

The SAP SuccessFactors Employee Central system provides a one-stop-shop for a global system of record.

In the following sections I'll review at a high level the features and functionalities of the Employee Central solution.

Globalization and Localization

The global capabilities, referred to as *globalization,* allow for storage of data related to specific countries, making the solution reliable for companies that require the ability to store global data, with variances across the countries using country-specific fields. These localizations are delivered providing legal compliance and reporting in seventy-one countries and thirty-five languages.

Localizations include country-specific fields, picklists, validations, rules, reports, and other features. This allows organizations to maintain a global template that is used across all countries and pair it with the ability to capture data and processes specific to certain geographies.

These localizations include national ID formats and data input validations, addresses, legal entity and job classifications, personal and employment information, reporting information and picklists and validations. Figure 1.35 shows how the system can support multiple address formats.

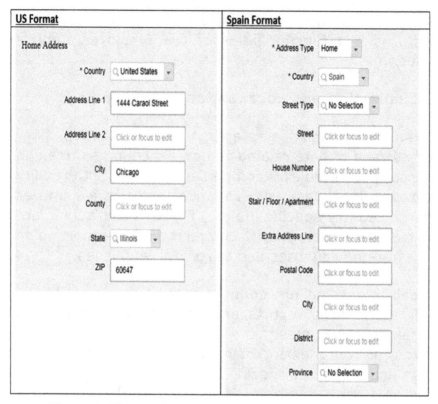

Figure 1.35: Localizations for Country-Specific Fields

Employee Central Basics

Objects and Data Models

Employee Central is built on effective-dated platform to allow for historical data viewing, tracking, and auditing. It is used to store organization, job, and pay information as well as the employee data, such as personal, job, or compensation details.

Organization information includes details associated with legal entities, business units, departments, divisions, and cost centers. The foundation objects also allow for storage of job and pay-related data including job codes, job families, pay ranges, pay grades, and geographic differentiators. Besides the storage of foundational data related to the organization, job, or pay structures, the solution also has a robust framework for storage of employee or user data.

The system design is best understood when broken out into the following areas:

- *Foundation Objects:* Used to populate employment objects. Examples include legal entity, departments, cost centers, pay ranges, pay grades, job codes, locations, and more. Figure 1.36 has a screenshot of a foundation object.

- *Person Objects:* Objects related to the person such as name, birthdate, address, and national ID.

- *Employment Objects:* Objects related to the employment like job information and compensation data.

- *Generic Objects*: includes additional capabilities built on the Metadata Framework (MDF) such as position management or time off.

- *User Objects:* The user account that is automatically created when an employee is hired populating the user data file (UDF).

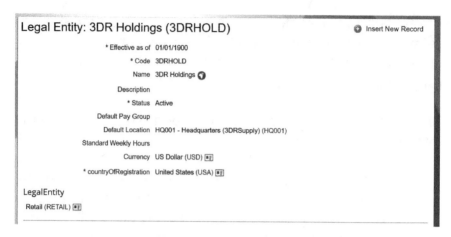

Figure 1.36: Legal Entity Foundation Object

Implementation Steps and Considerations

There are a number of very important considerations for an organization to make when implementing Employee Central. The journey to the cloud is not a short one, and while the Employee Central Implementation is similar to the other modules in the HCM suite, it's even more important when implementing this module to consider your current state analysis and data conversions.

- *Current State Assessment*: Take the time to assess your current processes, procedures, and data before the start of the project. This will ensure you have brought all of the points to the forefront reducing the risk for needing to add a field down the line that could cause a significant delay in the project.

- *Data Conversions*: Start gathering, reviewing, and cleansing your data prior to the start of the project. This will support an easier process when converting data into Employee Central, and more importantly, it will guide you in design decisions during the first iteration.

As you look at your current state assessment and data items, here are some of the areas to consider:

- Where does the data come from?

- Are there any forms that are completed for entry?

- Where does the data get entered?

- What type of employees does this change impact?

- Who needs to approve the action or change?

- What are the pain points associated with this process?

- How would you like to see this managed in the future?

- Are there any regional or country differences?

- What limitations exist that cause manual processing?

Events and Workflows

In an HRIS solution there is the need to capture the reasons for employee changes throughout their tenure. Employee Central has two options for capturing this. The first is to allow the person keying to select the event and the event reason. Alternatively, your organization can select to use event derivation, which will allow for automatic assignment of event and event reason based on the fields changed. This is an excellent and exciting way for organizations to reduce the need to train managers on the rules associated with specific change reasons and greatly improves reporting accuracy.

Workflows are the approval process that triggers from such changes. Once saved in the system an e-mail notification will be triggered to the approver, letting him or her know that he or she must take action. These changes do not hit the system until all approvals are complete. Figure 1.37 has a copy of a workflow reminder e-mail.

Figure 1.37: Workflow Reminder E-mail

There are various tools to allow the HR team to keep an eye on such transactions, such as admin alerts on the homepage or various reporting options.

Extensibility

In Employee Central, extensibility refers to the ability to take advantage of additional features. Figure 1.38 has the three areas that support extensibility.

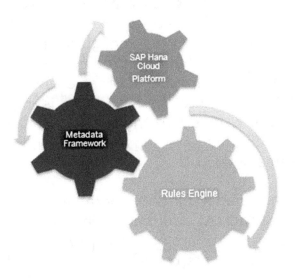

Figure 1.38: Extensible Capabilities

- *Metadata Framework (MDF):* Allow for building of custom objects. Currently, position management, time off, timesheets, global benefits, deductions, and advances are all built on the MDF platform.

- *Rules Engine*: allows for defaulting or validation of data that are triggered based on configuration options such as on initiate, change, or save of various forms. Functions include date and time, mathematical, and derivations as well as the ability for pop-ups to display various messages or warnings where supported.

- *SAP HANA:* ability to build custom applications that are displayed within the Employee Central feature.

Data Imports and Data Migration

The process to import data into Employee Central can, at times, seem tedious. There are over thirty CSV templates aligned with each of the tables, both foundation and employee, that will need to be populated and imported into the system in line with the import validation rules that are dependent on your configuration. Beyond the obvious frustrations that the process can cause, it is a reliable and furthermore affordable and easier way to import the data into the system. It is important to build appropriate time into the project to allow for the cleansing of existing data and manipulation to fit within the templates defined.

Another consideration for data migration is the historical loading of the data. As a common rule, bringing in history will always make a project more difficult. Generally speaking, many clients choose to bring in at least one year of history to allow for rehire and specific reporting needs. If using Workforce Analytics (WFA), you might consider bringing in at least two years of history. Whatever the decision it's important to discuss this with your Implementation partner during scoping and to ensure adequate time is built in to the project plan to support the migration.

Finally, consider your cutoff or freeze date. Ask the appropriate questions. Will you perform double entry? How long will users be unable to access the system? Which date will you use as the conversion date? Does the conversion date need to be clearly defined in the event reason for reporting purposes?

Time Management and Benefits

The time off feature within Employee Central allow employees and managers to manage workplace absences. This highly configurable module is independent of Employee Central allowing organizations to choose and customize the implementation. It allows for viewing balances, reporting, requesting time off, and approval workflows. Figure 1.39 has a screenshot of the time management panel.

Figure 1.39: Time Management Screen

Payroll time sheets allow for entering time worked in a weekly format. This feature is also highly configurable allowing organizations to build the rules associated with an employee's time such as overtime calculations that are viewable by the employee immediately upon entry. Built-in workflows allow managers to approve time at which point the data can be passed to a payroll system for processing and payout.

The global benefits feature allows organizations to maintain data beyond regular salary. Here organizations can manage benefits, benefit plans, and benefit enrollments. Types of benefits stored include allowances, reimbursements, pensions, company assets (car or cellphone, etc.), insurance, and more. Customers can choose to build these benefits custom or take advantage of the many predefined benefits that SAP SuccessFactors delivers.

Integration

Employee Central has many options for integration to on-premise solutions, both SAP and non-SAP systems, and there are a variety of tools available within these applications to allow for the management of both organization and employee data. I have covered Integration capabilities in more detail later in this chapter. Specific to Employee Central, however, there are additional prebuilt integrations that SAP SuccessFactors has built out to several third-party systems such as Kronos (time management), Aon Hewitt, Benefitfocus Thompsons (benefits), ADP, NGA (payroll). The use of these prebuilt integrations can reduce the cost and durations for the integration work. When using integrations, you do want to first look to see if there are any standard and prebuilt integrations that will work before starting to develop custom integrations.

Other Features

Employee Central is a maturing solution, and as such, a number of additional features are available, including global assignments, contingent workforce management, mass changes, concurrent employment, alternate cost distributions, and document generation, to name a few. As I will cover in more detail in Chapter 7, it is important for EC clients to stay abreast of new features that are included in the quarter releases. There are more updates for EC than any of the other modules.

In addition to new features introduced during the quarterly releases, there are a number of ways the system capabilities can be extended. Do discuss with your Implementation consultant if there are gaps in the system capabilities when reviewing the current and future state analysis so he or she can make recommendations to configure and deliver a solution that meets the needs of your company.

Platform Components

Most SAP SuccessFactors projects focus primarily on the individual modules being implemented. This approach overlooks the importance of the platform components and how they can impact the use of key features in the system. If platform components are not set up correctly, they can also have a negative impact on the roll out of additional modules in future phases. There are many platform components, and I would need another book to cover all of the capabilities in detail. I will cover them at a high level so you understand their function and potential impact on your implementation.

A platform component is essentially any configurable feature set that is used by multiple modules. In some training content and documentation, platform is often referred to as foundation elements. To avoid confusion with the foundation objects that are part of Employee Central, I will use the term *platform* to describe these shared features. Figure 1.40 is a graphical representation of the primary SAP SuccessFactors platform components.

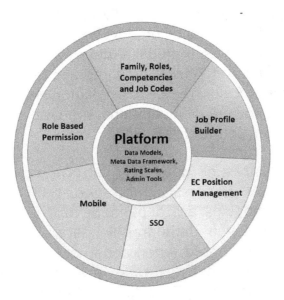

Figure 1.40: Platform Components

Some platform components are required for all clients, while some are optional. For example, Role-Based Permission, Data Model, Rating Scales, and Admin Tools are used by all clients regardless of whether they're implementing one or multiple modules. Features such as SSO and mobile are dependent on the client's internal capability and whether they have policy supporting the use of these features.

Other platform items such as Metadata Framework (MDF), Family and Roles, Competencies, Job Profile Builder, and Employee Central Position Management are not only dependent on the modules being implemented, but also on the client's need and their ability to define and maintain the data in these components.

For example, if a client does not have a defined competency model, they will only realize limited benefits from the implementation of Family and Roles. It would be better for them to focus on developing a competency model and then use their model to determine how best to configure Family and Roles. The same applies to the Job Profile Builder or Employee Central Position Management. If there are no formal job definitions, then little value can be derived from Job Profile Builder.

To visually represent how the platform components support the SAP SuccessFactors suite, refer to Figure 1.41. All the modules are dependent on the employee profile for employee demographic data. If Employee Central is implemented, then most of the employee demographic data in the profile is synchronized and updated from Employee Central. Whether the employee demographic information is coming from the employee profile or Employee Central would be totally transparent to the users.

From Figure 1.41, you can also see that the full HCM suite is impacted to some extent by the configuration of the platform components. As an example, with Role-Based Permission you can define permission access for all the modules including Employee Profile and Employee Central. Similarly, if Single Sign-On is used, all the modules in the HCM suite can be accessed once a user is authenticated.

Figure 1.41: Platform Components Supporting HCM Suite

As mentioned earlier, I will not cover these components in detail, but just highlight their importance. I strongly encourage you to spend the time up-front with your implementation partner to discuss how these components will affect your overall project. It's a lot easier than doing a retrofit in later phases of the project.

Family, Roles, Competencies, and Job Codes

This component is the legacy method to manage and link competencies with employees in the system. The Job Profile Builder, which we will cover later, is the replacement for Family and Roles. You can choose to use either one of these components, so you do have that flexibility, and for some clients Family and Roles will be a better option than Job Profile Builder.

There are several elements in this component, so before I get into the details, I'll define each of these elements.

Job families are the parent element and are used to group related job roles together. They are typically broad categories such levels or departments (HR, IT, Sales).

Job roles describe different types of jobs that share a common skill set. They can include a job description and a set of competencies. Each job family can have multiple job roles such sales manager, regional VP, etc.

Job codes are assigned to individual employees as part of the employee data file or Employee Central and are used to link the employee to a job role.

Competencies are a set of clearly defined and observable skills, abilities, behaviors, and knowledge required for successful performance in a role, position, level, or job.

These elements are connected together to create a conceptual hierarchical structure as represented in Figure 1.42. There are three fixed levels, but you can use these levels to represent any structure that makes the most sense for your organization. A small organization might have a single job family that represents the entire company. It would use the job role to represent the career type such as officers, mid-level leaders, and team members. It could use the job code to identify each unique position in the various career type. So at the officer level, one job code would represent CEO, and another would be used for an EVP, etc.

A slightly more complex structure can be created for a larger company where multiple job families can be used. In this case each job family would represent a department.

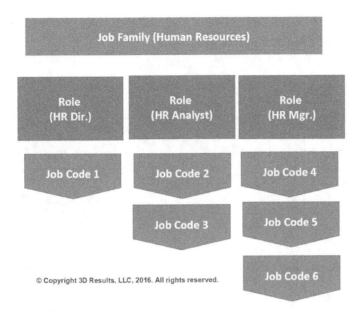

Figure 1.42: Hierarchical Structure of Job Family and Job Roles

In this example, the job family is Human Resources and the job roles are all the related roles. The HR manager role has multiple job codes associated with it, so multiple positions can be part of that role. The job code MUS830J is included in the employee record so the system knows which role to associate with the employee. These are just sample names. The client naming convention can be very comprehensive. For example, one client used each character in their job code to represent a value such as country, location, department, etc.

The job role, as highlighted in Figure 1.43, is what holds this structure together. It is not only the parent for the job codes; it is also where the position is actually defined. It can store a description and display it in the PM form. The role can also be associated with a talent pool. The key function of the job role is to define all the competencies that are relevant to this role. For each competency, the system also supports the capability to set an expected Rating and expected weight.

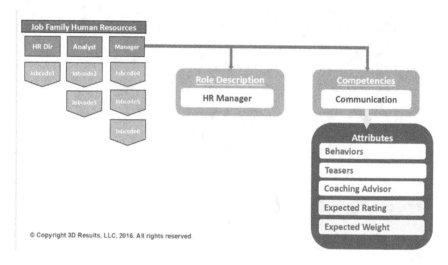

Figure 1.43: Role Attributes and Details

In Figure 1.43, you'll also notice that competencies are a significant part of this overall structure. Competencies are used across multiple modules within SAP SuccessFactors and can drive desired organizational values. They are often included in a performance review and can be used to assess an employee for a different position within the company or for areas of improvement in his or her current position. Competencies are also used as a standard benchmark so organizations can align, discuss, and measure talent consistently across all talent management processes and departments. Since they can be used in multiple ways, their use case can be different based on the module.

Before jumping into the use of competencies in specific module use, I'll review the competency structure shown in Figure 1.44. Each competency is assigned to a category and then each category is associated with a library. This means that you can have multiple competency libraries in the system. SAP SuccessFactors provides an excellent competency library with behaviors and a writing assistant that most clients use. However, if a client has developed their own competency library or is using a library from a third party, then it can be imported into the system.

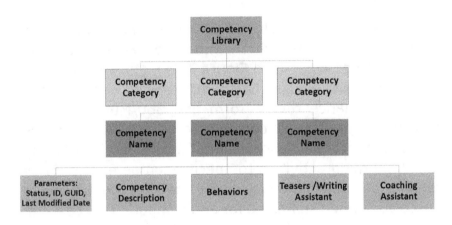

Figure 1.44: Competency Structure

The competency category is used to filter competencies. If there is a specific set of competencies that need to appear in a performance review form, they can be assigned to a category that gets included in the review form. This is a more dynamic approach since you are not hard coding individual competencies each time in each form.

The competency itself has multiple attributes that are defined in Figure 1.38. Behaviors can be a more granular definition of the competency, and there can be multiple behaviors assigned to one competency.

The writing assistant provides summary descriptions of what a behavior might look like at different levels of performance (needs improvement, meets expectations, exceeds expectations). The coaching advisor includes a range of coaching advice for each competency. The teasers and coaching advisor make it easy for users to rate themselves while providing constructive feedback since it provides example content and more detailed definitions.

Competency Use by Module

In the SAP SuccessFactors HCM suite, there are seven modules that can be directly impacted by the setup and use of competencies (see Figure 1.45). Most consultants and clients are aware that competencies can be used in a performance review form and 360/Multi-Rater. In addition, competencies can also be used in Recruiting as part of the interview assessment feature and in Learning to link learning activities to specific competencies. They can also be used in the Workforce Analytics module and in Succession when searching for and comparing successors.

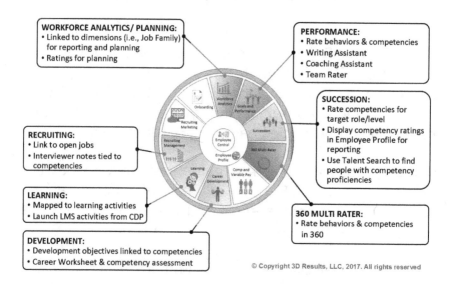

Figure 1.45: Where Competencies Are
Used in SAP SuccessFactors

Job Profile Builder

The Job Profile Builder is a platform feature in the SAP SuccessFactors HCM suite that provides a flexible and intuitive way to identify the complete elements of a job and share them with your users. It is meant to be a replacement for the legacy job description manager (JDM) so you can use this tool to control changes and updates to job profiles. Job Profile Builder or JPB for short is also a replacement for Families and Roles but provides more comprehensive features to allow customers to build complete job profiles with multiple content types.

JPB can be used with the skills management feature to allow customers to capture employees' skills within an employee profile portlet. There is an additional cost for the skills management feature, but it does come with prebuilt content that can make the setup and use easier.

One of the key decisions that you will need to make is whether to use Job Profile Builder or the legacy Family and Roles. There are some pros and cons on the user of either one of these components. Before we compare them in more detail, let's take a more granular review of the Job Profile Builder. Figure 1.46 has a view of how you can access JPB from the Employee Profile module and also shows some of the job elements.

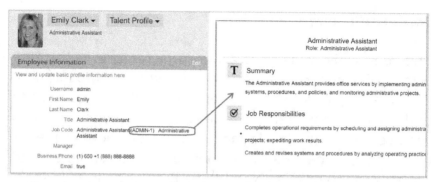

Figure 1.46: View of Job Profile from the Employee Profile

This sales director job profile that is displayed is associated with a job profile template. It is similar in concept to a goal plan template or a requisition form template. You can have as many templates as needed, and each of them can have different elements. Of course the more templates, the more complex and time-consuming it will be to maintain and use. Each template can then have one or multiple job profiles associated to it as reflected in Figure 1.47.

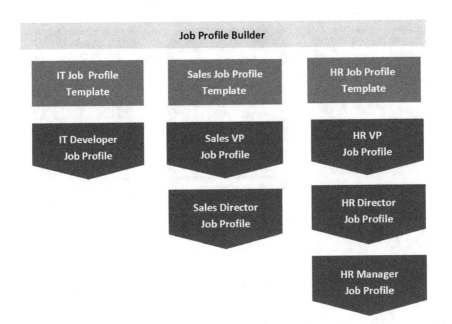

Figure 1.47: Relationship of Job Profiles to the JPB Templates

This flexibility to have different job templates allows you to tailor the information that will be captured in a job profile based on the positon requirements. So for example if there is no need to include physical requirements in a corporate job profile, but physical requirement is needed for field job profiles, then this can be supported. You will create one job template for corporate and exclude the physical requirements element and then create another job template for the field positions and include the physical requirements element. Figure 1.48 shows the elements of a job profile and also how Competencies can be associated with job role and job family.

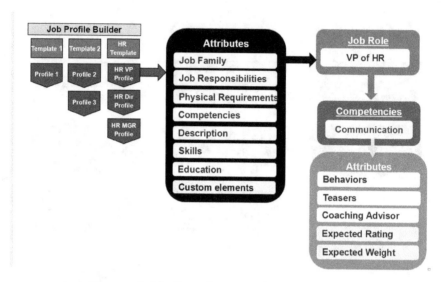

Figure 1.48: Detail View of a Job Profile

Since JPB also uses Family and Roles, at a minimum all the modules that use legacy Family and Roles will also use JPB. In additional, there will be some modules such as Recruiting, Career and Development Planning, Succession, and Employee Profile that are more tightly integrated with JPB and can use more if not all of the full JPP tool set. Figure 1.49 shows all the modules that are impacted by JPB.

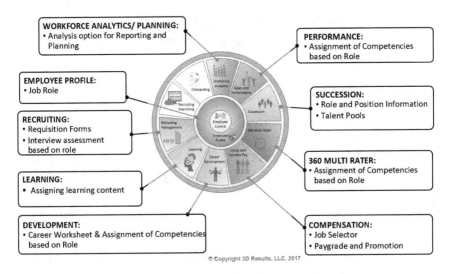

Figure 1.49: Modules Impacted by JPB

As mentioned earlier in this chapter, you can use either legacy Family and Roles or JPB, not both. The question then is which one you should use. There are some pros and cons that are highlighted below.

Benefits of using JPB:

- Job Profile Builder is the future direction of SuccessFactors platform for managing job family, roles, and competencies.

- It provides better (more advanced) integration options supporting Succession, Recruiting, and the Career Worksheet.

- Clients can define and use more granular definition for Job Profiles.

- Job Profile can be combined with Skills Management to get a more comprehensive picture of an employee capabilities.

- It can be used as a repository to store job definition elements until they are needed to create the job profiles.

Disadvantages of using JPB:

- It requires conversion of existing job family, role, and competency set up. There is a migration path available, but some manual rework might still be needed.

- JPB current supports the use of the writing assistant and coaching advisor (WACA), but modifications to WACA are very limited.

- Competency descriptions are limited to text. Tables are not supported, and HTML rendering is inconsistent.

- Behaviors cannot be mapped at the role level. When a competency is assigned to a role, all the behaviors associated with that competency are also associated with the role. In the legacy Family and Roles, the option existed to selectively identify which behaviors from a competency should be associated with a role.

- Does not support hidden competencies.

- Can only map roles by job code field (custom field mapping not supported).

- This is a more complex construct to create and maintain versus the legacy Family and Roles and will require much more effort and skills from the client personnel.

The decision about whether to use legacy Family and Roles or Job Profile Builder will vary for each client. It is key to have this discussion early in the implementation so adequate time can be allocated to the setup and use of either one of these features.

EC Position Management

EC Position Management is used to manage the creation and maintenance of positions in Employee Central. This position object is used to autopopulate the employee records with organization and position attributes. Position data is also used by Succession, Recruiting, and Employee Central, so it allows you to have a standard definition of a position in multiple parts of the system. This allows for consistency in the position data that is used to populate the employee record, improving data quality and efficiency.

EC Position Management is an MDF (Metadata Framework) construct so the position attributes can be customized based on your requirements. Business rules can be used to manage changes to the position object, and RBP (Role-Based Permission) is used to control access to the positon objects.

There are two main components, the position organization chart and the position definition. The position organization chart is a hierarchical representation of the position structure that shows the relationships between positions. It is similar in concept to the company organization chart. A sample position organization chart is displayed in Figure 1.50.

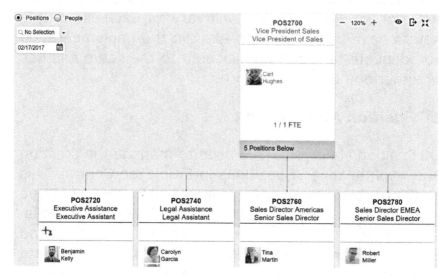

Figure 1.50: Position Organization Chart

The position definition is the MDF object that defines the fields and attributes of the position. This stores the data that will autopopulate the employee record. From the positon organization chart, you can drill down to view the detail attributes of the position. Figure 1.51 shows the details for the vice president position.

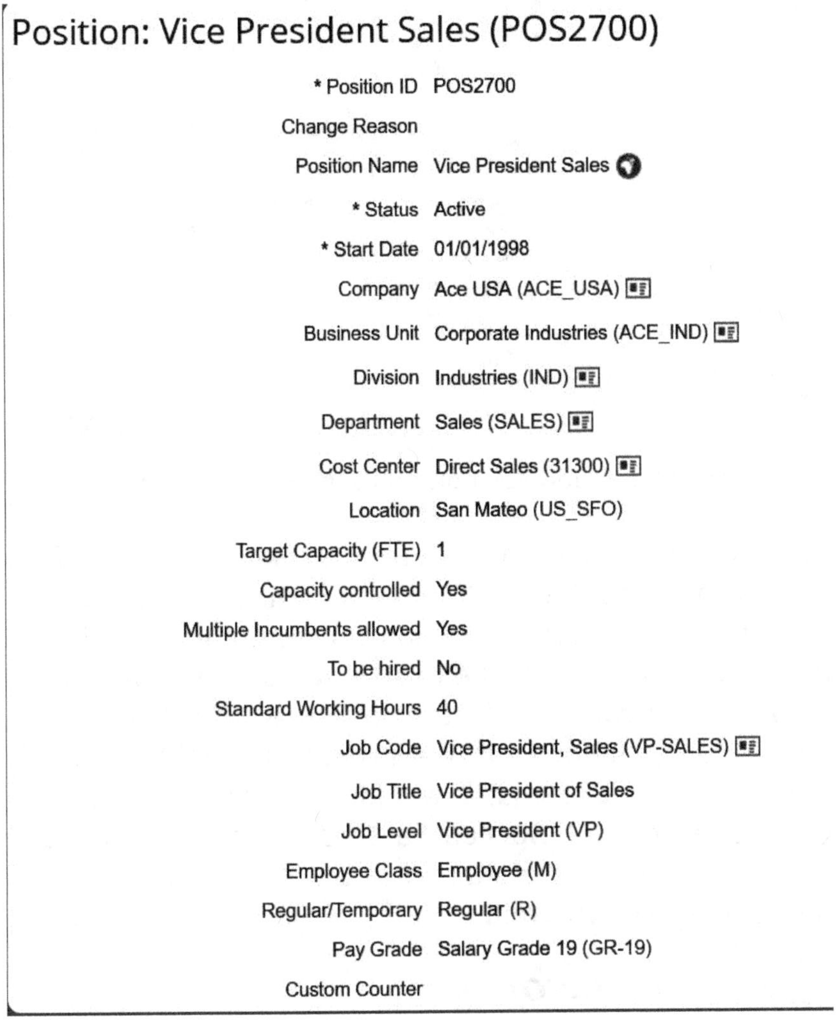

Position: Vice President Sales (POS2700)

* Position ID	POS2700
Change Reason	
Position Name	Vice President Sales 🌐
* Status	Active
* Start Date	01/01/1998
Company	Ace USA (ACE_USA) ▦
Business Unit	Corporate Industries (ACE_IND) ▦
Division	Industries (IND) ▦
Department	Sales (SALES) ▦
Cost Center	Direct Sales (31300) ▦
Location	San Mateo (US_SFO)
Target Capacity (FTE)	1
Capacity controlled	Yes
Multiple Incumbents allowed	Yes
To be hired	No
Standard Working Hours	40
Job Code	Vice President, Sales (VP-SALES) ▦
Job Title	Vice President of Sales
Job Level	Vice President (VP)
Employee Class	Employee (M)
Regular/Temporary	Regular (R)
Pay Grade	Salary Grade 19 (GR-19)
Custom Counter	

Figure 1.51: Position Details

The Job Profile Builder that I have listed as another platform component also can use job positions if they are defined in the system. In Figure 1.52 below, I have a screenshot of the Job Profile Builder screen where you can associate the job roles with job families and the job positions with job roles.

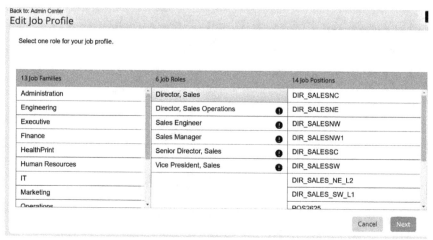

Figure 1.52: Job Position and Job Profile Builder

Since Position Management can be used in Job Profile Builder, Employee Central, Succession, and Recruiting, do discuss with your implementation consultant if it should be implemented. It can only be used if you are implementing Employee Central, but if you have granular definitions for your job positions or this information is needed in payroll or other systems, it would be a good idea to use this feature in SAP SuccessFactors. This way position information can be maintained in Employee Central and fed to the other systems.

Single Sign-On (SSO)

Single Sign-On provides a secure and seamless method for a user to login to a system without entering his or her credentials. For example, the normal process for a user to log in to the SAP SuccessFactors or other cloud-based application is for the user to click on a URL and the browser will take him or her to the login page. Here the user will enter his or her username and password to get access to the system. When using SSO, the user only needs to click on the URL; there is no need to enter any username or password. This makes it a lot easier to access the system, so SSO is used by many companies.

The way SSO works is that behind the scenes there is some authentication that occurs whenever you click on the URL link to connect to SAP SuccessFactors. It is totally transparent to the user, but once this validation is completed, the browser then has some additional information about your account, and it uses this information to submit a request to SAP SuccessFactors. When SAP SuccessFactors receives this information, it looks at it and if all the parameters can be validated, it gives you access to the system. In order for this background authentication to occur on your side it does require that your company have the architecture in place to support SSO. The type of system setup needed will be dependent on the method of SSO that is being used. Fortunately, SAP SuccessFactors supports multiple SSO methods.

These SSO methods can be classified into two main groups. The first is token-based and the second is called SAML, which is an abbreviation for Security Assertion Markup Language. SAML is based on a predefined standard for exchanging authentication information between systems. If your company is already using SAML for SSO with other applications, then most likely that is what will be used for SSO to SAP SuccessFactors since it is the most secure SSO option supported by SAP SuccessFactors.

In the event that SAML is not being used by your company, then your technical folks can look at one of the token-based options. Here is the list of token-based SSO options currently supported by SAP SuccessFactors:

- token only

- MD5

- MD5 with Base64 Encoded user name

- SHA-1

- DES

- 3DES

Clients can develop their own application to support token-based SSO, so it provides a bit more flexibility on how it can be implemented. If SSO is not currently used in your company, it will take some time to decide on which type of SSO to use and then also to build out the infrastructure to support SSO. For this reason, it would be best to implement SSO after the initial launch of the SAP SuccessFactors system if the architecture is not in place early in the implementation.

The process I just covered is for connecting to SAP SuccessFactors from your company network. But what happens if you are in the SuccessFactors HCM suite and want to use SSO to connect to a third-party system? This is also supported. SAP SuccessFactors has built out SSO capability from their system to some third-party vendors such as ADP, Benefit Focus, and Kronos. An even if you have a third-party vendor that SAP SuccessFactors is not currently configured to support SSO, it is still possible to set up SSO to other systems, though there will be some limitations to the SSO methods and access points from within SAP SuccessFactors that can be used.

The configuration for custom SSO will be dependent on the type of SSO method being used for connecting to the other system. If you are using the token-only option, there is a feature called custom navigation links with dynamic parameters that can be used to support this type of SSO to another system. If SAML is used, then SSO to another system from SuccessFactors is still possible, but the request will need to be routed through your local identity provider so the user can be authenticated or look at the Jam module SSO capabilities for connecting with other systems.

In additional to using SSO to connect to a third-party system from SAP SuccessFactors, there are two other questions that are often asked by clients when implementing SSO for SAP SuccessFactors. One is whether an integration is needed with the client directory system, and the other is if the same password that is used on their internal network is required by SAP SuccessFactors. The answer to both of these questions is no.

There is no direct integration required for SSO. All of the SSO options supported by SAP SuccessFactors do not require an integration between your company directory system and SAP SuccessFactors. The two systems do not directly interact with each other. Your internet browser is the medium that carries the information from one system to the other. Figure 1.53 has a simplified graphical representation of how this process will work.

Figure 1.53: High-Level SSO Process

As you can see from the above diagram the interaction with the user and the identity management system where the user authentication occurs is typically done via a browser. Some users may have a direct login to their identity management system if it exists within their firewall, but I am finding a lot of SAP SuccessFactors clients are now using cloud-based identity management services such as Okta. Regardless, the actual submission of the request to SAP SuccessFactors is via a browser, so there is no direct integration involved.

On the second question relating to passwords, these are definitely required for the traditional login method, and they are also required for some SSO options. For other SSO options like SAML or MD5 with Base 64, a password is not required. However, even if a password is required for security purposes, it should not be the same password that is used for authentication to the client internal systems. SAP SuccessFactors does have a capability to generate random values for password for new hires, which is the option most clients will use when initially seeding the system with user data.

Partial SSO

Before I close out this chapter, I do want to cover a feature called partial SSO that is supported by SAP SuccessFactors. This feature allows some users to login with passwords while the rest of users are forced to use SSO. A user is either classified as a SSO user or a password user. It does not allow any one user to login both ways. This means that you cannot use SSO in the office and password login when you are home. If you are setup as a SSO user, then regardless of where and when you are connecting to the SAP SuccessFactors system, you will be forced to login via SSO. This does not include mobile device access for which SAP SuccessFactors has a separate security protocol that is used to control access.

This feature comes in handy for organizations that want to give access to SuccessFactors to users who are not in the directory system that is used by their SSO application. Typically, we find that companies that use contract and temporary workers like this feature. It is also used for organizations that make their Learning modules accessible to their vendors and clients since these individuals will also likely not be in the directory system.

If partial SSO is a feature that you would like to have implemented, then inform your implementation consultant early in the project since there are some additional configurations needed to support this feature. It also requires a change to the employee demographic file, which is covered later in the section. The team that is creating this file will also need to be able to identify which users are SSO users and those who will be password users.

SSO can be a fairly complex and technical topic. In this chapter, I have distilled this information to keep it very high level. It is one of the areas in the deployment of SAP SuccessFactors that you definitely want to involve your technical team. It is not only so they can implement it for you, but because your organization may have security protocols and a chief security officer who would want to review and approve the use of SSO for any external system. It is better to engage the right personnel from your internal team early in the project so they can complete all their reviews and approvals way before you are ready to complete testing or launching of the system to your end users.

Mobile

The SAP SuccessFactors suite can also be accessed from a tablet or a smartphone in addition to desktops and laptops. This allows you to extend the reach and use of the SuccessFactors system and realize more benefits from the implementation. There are multiple activities that can be performed on the mobile devices from simple viewing of data such as organization chart and directory search to more complex activities such as job requisitions approvals and requesting time off.

If you are going to access the system using mobile devices, then this decision needs to be communicated to the implementation team early in the design process. The reason is that the use of mobile devices requires some additional system configuration and setup. Also you will have to decide on which of the available functions in the system will be accessible by mobile device and ensure that they are in compliance with your corporate mobile device policy.

Ideally, your company should already have a policy in place concerning the use of mobile devices to access corporate data. This will provide some guidance on what information you can make accessible via mobile devices. In the SAP SuccessFactors suite the type of data and the actions that can be performed using mobile devices vary by modules. Your implementation consultants can provide the details based on the module you are implementing. Figure 1.54 has a high-level summary of the functions that can be accomplished on a mobile device.

Employee	Manager	Executive	IT
Productive Employees	*Better People Managers*	*Insightful Leaders*	*Agile Enablers*
• Employee Profile • Learning • Goal Management • Social Collaboration • Time Off • QuickGuides • New Hire Onboarding	• To-dos & Approvals • Recruiting workflows • Organizational Chart • Touchbase • Who's in the meeting • Manager Onboarding • Manager Cockpit	• Workforce Analytics • Headlines • Notifications • SuccessFactors Presentations (Beta)	• Role-based Permissions • Fast Deployment tools • Secure Activation • High-security Encryption • Remote deactivation • Multiprofile switcher

Figure 1.54: Mobile Capabilities

As you decide on which functions to make accessible on mobile device, do consider user adoption and ease of use. We will cover guiding principles later in Chapter 2, but a good rule of the thumb is not to use a feature because it is available, but because it will simplify the work for your employees and can provide a tangible benefit. This applies to both mobile devises and features in the SAP SuccessFactors suite as a whole.

Role-Based Permission (RBP)

The security framework that SAP SuccessFactors uses to control end-user access and secure data in the system is called Role-Based Permission or RBP. This is a very comprehensive framework that allows you to have very granular permission control such as at the field level. It is fairly easy to use once you figure out how it works. In most projects, the implementation consultants will create a few standard roles and provide some training so the client system administrators so they can continue to create and maintain RBP.

There are two main components to RBP. The first is a called a permission group and the second is called a permission role. The group is comprised of a list of individuals. This is often a dynamic list since you can use filters such as division, department, and other custom values to determine who should be in the group. As employees get added or their filter values are modified, they will be dynamically added or removed from the group. For example, you can create a group called Talent by simply including a filter to limit the group list to employees whose department is Talent Management. Figure 1.55 shows the screen from the Admin Tool section of the system where you will create the various permission groups. In this case the use of a single filter for Talent Management restricts the group to the thirty-six employees who are currently in this department.

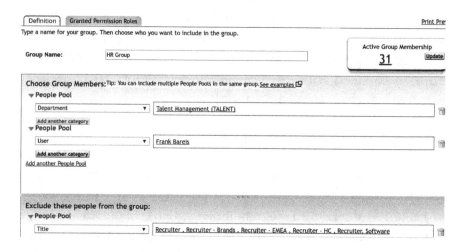

Figure 1.55: Creating a Permission Group

A permission role on the other hand defines what data can be accessed and the types of transactions that can be effected on the data. For example, you can create a manager role that provides access to performance and talent information for their direct reports. The manager role can also be permissioned to allow a manager to change an employee location or to give them a promotion. Figure 1.56 shows how you can setup the permission that will be associated with a role. In this case any user assigned to this role will have the ability to update the access permission for the development plans.

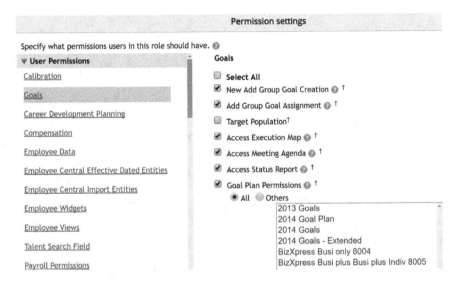

Figure 1.56: Permission Role

Once these two components have been created, they are then combined together to complete the setup for a particular permission group. In Figure 1.57, we have the relationship between the group and the role.

Figure 1.57 Relationship between the Groups and Roles

Both the granted users (USA HR) and the target users (USA employees) are permission groups. The permissions that are available to the USA HR group are based on the authority assigned to the HR role. This diagram is saying that all the users in the USA HR group will have access and ability to change data for anyone in the USA employees group based on the authority that is granted to them in the HR role.

You can make as many permission groups and roles as needed and then combine them to complete the permission assignment. The more groups and roles that are created though will make it more difficult to maintain and manage, so it is best to optimize the number of roles and be as efficient as possible. Most often this responsibility is assigned to the primary SAP SuccessFactors system administrator and their backup so it is centralized. This helps to ensure that there are not duplicate roles and groups being created. Since RBP controls access to user data, it is a good idea that this responsibility be restricted to a select few employees in the company.

Core Platform

In this section I have included all the other administrative and configurable elements that impact multiple modules.

Your implementation consultant will review these elements during the implementation. SAP SuccessFactors also provides administration training classes so your team can become more familiar with the capabilities and the impact these elements can have on the system. I will provide a brief overview so you can have an idea.

- Succession Data Model

- Metadata Framework (MDF)

- Admin Tools

- Rating Scales

The Data Model is a configuration template that determines the type of employee demographic data, Succession, Compensation and Employee Central information that will be stored in the system. This configuration will be impacted by the client-specific requirements around data and business rules. There are different types of data models that are used by the system, especially when implementing Employee Central. The one I am referring to here is called the Succession data model and is what drives the fields that can be seen in the employee profile. This Succession data model is needed for all SuccessFactors implementation since at a minimum a client would need to have their employee demographic data in the system in order to launch any of the modules.

Metadata Framework is a platform component that is used to add flexibility and extensibility to the system. You can define database objects and create relationships, business rules, and workflows. It is used primarily in Employee Central to build out new panels, but can also be used to support Succession. Recruiting currently does not use MDF, but there is some capability to sync some

MDF objects with legacy features in Recruiting to support position management. This is covered in more detail in EC Position Management.

Admin Tools refer to a comprehensive administration panel where an administrator will have access to a host of system features. They can make simple changes such as password resets and employee permission changes to more complex configuration of the system such as form template changes and home page setup. This is a very robust set of tools, so it is good idea to identify your system administrators early in the project so they can start to learn how to use the Admin Tool and be better able to support the system once it is live.

Rating Scales are used across many of the modules in the HCM suite and serve varying purposes in the system, so companies should consider the definition of their ratings scales and how they are planning to use it. For example, the use of a rating in the annual performance review process is well-documented, and this rating serves as the foundation for the some of the discussion between the manager and the employee. It is also used to stack rank and compare employee performance and in Succession, Employee Profile, and even in Compensation. This is however only one Rating Scale; there are many other Rating Scales used in the system.

The below diagram shows where Rating Scales are used in SAP SuccessFactors. Rating Scales are also used in Calibration and Job Profile Builder. Since these are not independent modules, I have not listed them in the above diagram. Calibration will inherit the rating scale from the primary module where the data is being sourced, such as Performance or Succession. In Job Profile Builder, there is a five-point skills assessment rating scale that is used. This is predefined by Sap SuccessFactors and not configurable in the system.

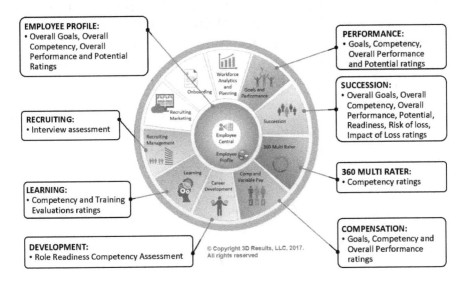

EMPLOYEE PROFILE:
• Overall Goals, Overall Competency, Overall Performance and Potential Ratings

RECRUITING:
• Interview assessment

LEARNING:
• Competency and Training Evaluations ratings

DEVELOPMENT:
• Role Readiness Competency Assessment

PERFORMANCE:
• Goals, Competency, Overall Performance and Potential ratings

SUCCESSION:
• Overall Goals, Overall Competency, Overall Performance, Potential, Readiness, Risk of loss, Impact of Loss ratings

360 MULTI RATER:
• Competency ratings

COMPENSATION:
• Goals, Competency and Overall Performance ratings

Figure 1.58: Utilization of Rating Scales

You can use different Rating Scales in many parts of the system, but where possible do try to standardize on the same rating scale or at least use the same number of values in the rating scale. If not, the system will attempt to normalize the values, which can be confusing to your user as they will see different values for their rating in various parts of the system. For example, if the overall competency rating uses a five-point rating scale in a performance review form and this information is presented in a succession nine-box (also called matrix grid), which uses a three-point scale, then the system will normalize the five-point scale to a three-point value in order to display in the nine-box. The user will then see a higher overall competency rating in the performance review form than what they will see in the nine-box.

In this section, I have covered the platform components in more detail to show the versatility and impact across multiple modules. Platform components allow for the standardization and use of the same object definition across the SAP SuccessFactors HCM suite, so investing time up-front in the implementation to understand their capabilities and reach will significantly impact project success.

Data Requirements and Integration

This topic is one that will often keep your IT department awake at night, especially if they are responsible for feeding the data to support the SAP SuccessFactors implementation. And it certainly does not help when they are told during the kick-off meeting that the data requirements will be finalized during the design sessions. Further aggravating the situation is that the time difference between design and build can be a month or less, which does not provide adequate time for your IT department to create their programs to generate the data.

There are a couple of things that can be done to alleviate these concerns. SAP SuccessFactors does have several options for importing and exporting data out of the system. And if this will be a substantial effort, you can work with your implementation partner or a third-party firm to get other consultants to supplement your internal technical team. Also, since not all of the data is required at the same time, there are some approaches like the one outlined below that can be used to make this a more manageable activity.

Employee Demographic Data

When looking at data requirements, I recommend a two-pronged approach that considers employee demographic data first and module-specific data second. There is a standard layout for the employee demographic data that can be provided to your technical team early in the implementation process. The actual field values change from one client to another, and there are fifteen custom fields that you can populate based on your design configuration. What's important is that the layout remains the same and the fields that you are not using can remain empty.

The standard layout also provides descriptions as to what each field (other than the custom fields) should contain—in particular, the required fields. This will give your IT team a head start in building out their programs to provide the demographic feed. Since this is the seed data that is used by all the modules, it is the first and most critical set of information required by the implementation team. Getting started on the demographic data early will help your project schedule stay on track. Figure 1.58 shows some of the field definitions that exist for the fifty-one fields in the employee demographic feed.

Col #	Header 1	Required (R) Optional (O)	Data Type	MaxChar	Instruction / Guidelines / Description
1	STATUS	R	String	32	Must be first field in the file Used to define employee status. Can be set to either "ACTIVE" or "INACTIVE". When first loaded, all users are to be set to "Active" BEST PRACTICE: It is not recommended to load all previously "Inactive" employees
2	USERID	R	String	100	Must be second field in the file Permanent, Unique ID for employee. Key to the system and must be set up as one word (one string, no spaces). This value is visible in a variety of places to all end users. Accordingly, USERID should not contain data that is considered confidential, such as social security number. This value must be passed with each employee data record during each upload. Values should be unique and one which would not change during the course of employee tenure. USER ID's cannot be recycled or reused.
3	USERNAME	R	String	100	Log in for users of the system. Can be any value desired by client. Usernames must be unique, but can be changed (ex: if marriage results in a name change, USERID would remain the same, but the USERNAME could be changed).

Figure 1.59: Field Layout for the Employee Demographic Feed

From an integration perspective, this information can be sent to the SuccessFactors secure FTP server where there is a standard batch process that can be scheduled to run daily to upload the information. Figure 1.59 is a data flow diagram representing this process (when Employee Central is not being implemented). This information will be uploaded into the employee profile. Those modules that were organically developed by SAP SuccessFactors (such as Performance and Development) are tightly integrated with the employee profile and will automatically be updated with this information. For modules that were acquired (such as WFA and Learning), there are synchronization jobs that your consultant can set up, but the overall integration process for the employee demographic data will be seamless across all the modules once it is uploaded into the employee profile.

Most clients will use CSV files and the SFTP server to send the data over to SAP SuccessFactors, but there is another option. SAP SuccessFactors has provided an application program interface (API) that can be used by another program or integration platform to make the updates directly into the employee profile. Based on your organization's integration policies and capabilities, they may have a preference of integration methods.

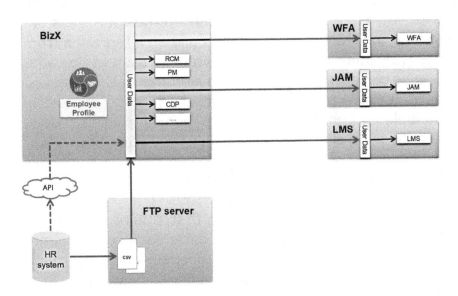

Figure 1.60: Employee Demographic Data
Flow (without Employee Central)

Module-Specific Data Requirements

Module-specific data requires a different approach that relies on the design process to determine what information will be required. For example, if it was decided during the design process for Succession that the organization wants to search on training courses taken by employees, then course completion data would need to be fed from the Learning module into the employee profile. Some requirements will be known early in the implementation process, but the field level requirements will only be uncovered during the design process.

Another consideration in the case of multiple systems is which one will be the system of record. If an on-premise system is the system of record, then a full or incremental nightly feed (using the previous example of training courses) of updated course completion data will be needed. If SAP SuccessFactors is the system of record, then you'll need to migrate data from the source system to feed the HCM suite and then build an integration from the HCM suite back to the on-premise system so the it is kept current. This entails extra effort, so there should be a decision as to whether the data needs to exist in multiple systems or if all your reporting can be done directly from the SAP SuccessFactors suite.

For a comprehensive view of all the potential data requirements, refer to Figure 1.60. The modules are shown across the top, and the first column contains a brief description of the type of data. Each row does not equate to a single file or record type. For example, trend, background, foundation data will typically be several record types and files based on the how the system is designed.

This does look a bit overwhelming, but much will depend on which modules are being implemented and the design configuration. Organizations with more talent management systems and processes will have more data, while those just starting to automate talent management processes will have a reduced effort for data migration and integration.

	RM	RMK	JAM	EC	LMS	EP	GM	PM	360	TR	CALI	COMP	VP	CDP	SM	WFA	WFP
Employee Data	X	X	X	X	X	X	X	X	X	X	X	X	X	X	X	X	X
Compensation Data				X		X					X	X	X			X	X
Variable Pay Data				X		X								X		X	X
Personal Data (Org Data)	X				X	X		X	X						X	X	X
Background Data	X				X		X					X			X	X	X
Trend Data (Ratings)					X		X							X	X	X	X
Competency Data	X				X	X	X		X	X	x	X		X	X	X	X
Families, Roles and Jobcodes	X				X			X	X				X	X	X	X	X
Position Data	X				X										X	X	X
Goal Data, Imported Goals, Library							X		X					X		X	X
Question Library	X															X	X
Learning Activities			X		X	X	X	X	X					X	X	X	X
Picklists	X				X		X						X		X	X	X
Foundation data				X													

Figure 1.61: Data Requirements by Module

The Employee Central module has a very different set of data requirements. The last row of Figure 1.60 has a reference to foundation data. This is all the organizational data that will be needed, such as department, division, job code, etc. The employee data requirement shown in the first row also refers to Employee Central. Typically, you will need to seed employee data into Employee Central from another system. Most clients will either do this from their on-premise system or their payroll system. Include adequate time for this effort as there will be transformation changes that need to be applied before the data can be uploaded into Employee Central. Data quality may also cause some delays; if you have missing, incorrect, or outdated information, the data load will fail and need to be corrected.

As mentioned, the subject of data requirements is a critical one for your IT team or whoever is responsible for providing this service during the implementation. Plan accordingly and include adequate time in the project for these activities.

Integration Capabilities

I have covered the type of data that is used by the system but not the tools that can be used to get data in and out of the system. This is also a topic of interest for your IT team since they are typically responsible for building the integrations. It is still a good idea for you to be familiar with the tools that support these integrations, in particular those for data extraction and reporting, since some of these activities will be owned by the system administrators. I will cover reporting in the next chapter, so for the remainder of this chapter the focus will be on integration tools.

Unlike on-premise systems, the database used to store your information is not directly accessible to you or your consultants. You can only access data through specific tools and pathways that have been developed for this purpose. Figure 1.61 is a graphical representation of these tools.

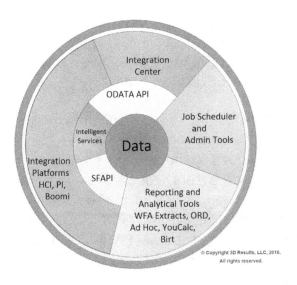

Figure 1.62: Data Extraction and Integration Capabilities

SAP SuccessFactors supports both SOAP-based API (SFAPI) and REST-based API (ODATA). Recently, SAP has indicated that SFAPI is legacy API technology so you should first look to use OData APIs when building any integrations. These APIs can be used to extract information from and update information in the system. Most clients will use these APIs via an integration platform such as Dell Boomi AtomSphere, SAP HANA Cloud Integration (HCI), SAP Platform Integration (PI), or with custom development applications.

Intelligent services are a set of automated services that get triggered by a specific change to user data in the system. For example, a French employee is transferring from Hong Kong to the New York office. This relocation event will trigger multiple actions in the system, such as a change in manager, address change, compensation location differential changes, and in-progress transactions that need to be rerouted to the new employee filling the Hong Kong role (onboarding, performance review, etc.). These actions can be internal and external to the system. This is a relatively new feature that can also be used to support real-time integration, and it has some tremendous potential, so I will cover this in a bit more detail at the end of this chapter.

The integration center is a recent offering that supports data joins, formatting, and some transformation capabilities. Reporting tools such as online report designer (ORD), Ad Hoc reporting, YouCalc, and Birt are primarily for generating reports. All of these (with the exception of Birt) are proprietary to SAP SuccessFactors. The Workforce Analytics module (WFA) also has the ability to pull data from SAP SuccessFactors for reporting and analytics.

The job scheduler is a very mature product and can be used to schedule both uploads and downloads from the system. The scheduler has access to over one hundred predefined batch jobs that can be used for integrations. Their usage will depend on the type of data, module, updates, or exports, and whether it is a manual execution or a scheduled run. One interesting aspect of the scheduler is that you can set up a custom ad hoc report job to run automatically, so it is not limited to predefined batch jobs delivered by SAP SuccessFactors but can also be used for scheduling custom report executions.

Depending on the size and complexity of the project, integration can be a significant part of the overall project scope. For smaller implementations, the native extraction tools such as the integration center, job scheduler, admin tools extract, and reporting output may be sufficient. If the implementation involves Recruiting or Employee Central, or there is a need to implement custom logic or triggering actions in real time to other external systems, then the use of the API and a more robust implementation platform such as HCI or Boomi will be required.

An integration platform can also be used for automating and compressing a business process. For example, when hiring candidates using the Recruiting module, compressing the time it takes to complete the process lowers costs and reduces risk. This capability is also known as auto progression and is essentially a workflow that automatically routes candidates through the recruiting process based on certain triggers or outcomes. For example, if an application is updated to reflect that the candidate failed a background check, then that person is automatically dispositioned and removed from the queue for that application. If the application is updated to reflect that the candidate passed the background test, then the person is automatically moved to the next step in the hiring process. There are multiple ways this concept can be used to result in a significant reduction in cycle times. You can even automatically bypass some steps in the process if they're not relevant to a particular applicant or position.

Another benefit of using an integration platform is to extend the system's capabilities. By building a seamless integration with other vendors, you can offer a service that is not natively available in your SaaS solution. This includes benefits, payroll, background verifications, and assessments. The objective is to summarize all the results and display them in one system so the user only has to go into that single system to complete an action. From a user perspective, even if there are two different systems and vendors, it will appear as a single integrated solution. An integration platform makes it easier to integrate with other cloud-based systems.

Figure 1.63 shows a screenshot of how SAP SuccessFactors can be extended to support background screening. As you can see, all the key information exists in SAP SuccessFactors for the recruiter to process. There is also a link to the detailed report.

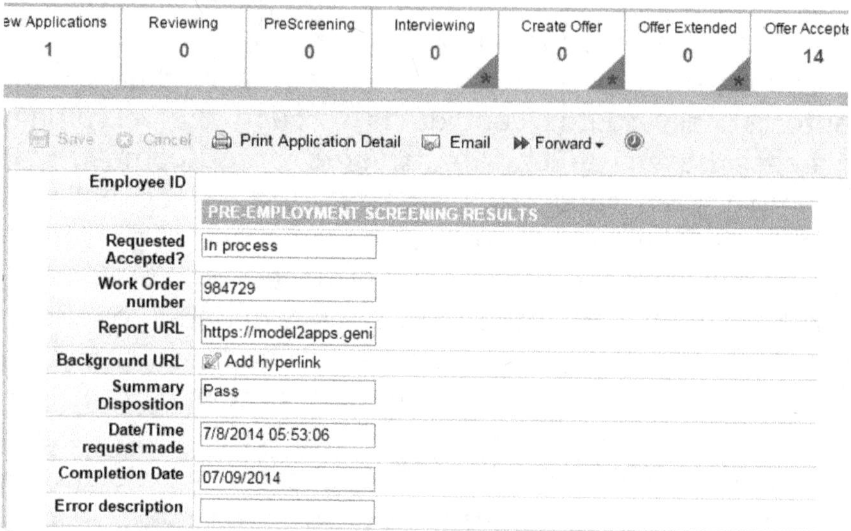

ew Applications	Reviewing	PreScreening	Interviewing	Create Offer	Offer Extended	Offer Accepte
1	0	0	0	0	0	14

💾 Save ❌ Cancel 🖨 Print Application Detail 📧 Email ▶▶ Forward ▾ ◉

Employee ID	
PRE-EMPLOYMENT SCREENING RESULTS	
Requested Accepted?	In process
Work Order number	984729
Report URL	https://model2apps.geni
Background URL	🔗 Add hyperlink
Summary Disposition	Pass
Date/Time request made	7/8/2014 05:53:06
Completion Date	07/09/2014
Error description	

Figure 1.63: Background Screening Summary
Information in SAP SuccessFactors

An additional benefit is that this extensibility can help the recruiter focus on the top candidates. They can sort the list of candidates to quickly identify those who have passed the background checks. Some clients also combine an integration with auto progression. For example, if the candidate has failed the background check, that candidate can be automatically dispositioned. If the candidate passes the background check, then he or she will be automatically progressed to the next step in the process.

If you are an existing SAP HCM client, then there are other tools that are available to integrate the SAP On-Premise system with your SuccessFactors cloud solution.

The key is to decide early in the implementation if the native integration tools provided by SAP SuccessFactors will support your needs or if a more robust integration platform is needed. This will give you more time to decide which implementation tool to use.

Intelligent Services

Intelligent Services is a relatively new product feature that can trigger updates in both SuccessFactors and externals systems when predefined event occurs. The primary use case is to automate dependent tasks that can be triggered by an HR Event so the user does not need to worry about downstream implications caused for example by placing an employee on long term disability when that employee is listed as a hiring manager on several open requisitions.

A secondary benefit is that Intelligent Services supports real-time integrations with the SAP SuccessFactors modules and external systems. Prior to Intelligent Services, all outgoing processes were batched and scheduled, but now updates in the system can trigger both internal and external processes. This is done using alerts or notifications from one module or component to another. Here is how it works:

1) A user updates information in the system;

2) This change triggers an event notification;

3) A dependent system subscribing to the event receive notification when the event is triggered; and

4) The subscribing system then executes its own process once it receives the notification.

In Figure 1.64, I have a graphical representation of how this feature works.

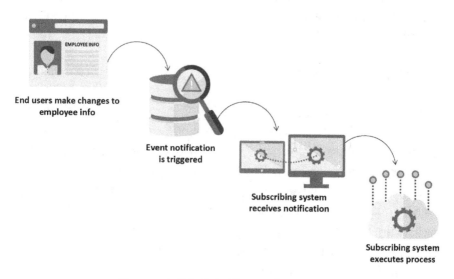

Figure 1.64: Intelligent Services

One key thing to note is not all data changes will trigger an event. Each event notification first needs to be released by SAP SuccessFactors before they can be used. With each quarterly release, there have been many new event notifications that are made available for consumption by both internal and external processes. Currently most of the event notifications supported are in the Employee Central module and are triggered when there is a new hire, termination, job change, and leave of absence, but as you can see from the list below, event notifications are now supported in many of the other modules as well.

Intelligent Service Events

- Recruiting
- Onboarding and Offboarding
- Hiring
- Job Changes
- Talent Management
- Continuous Performance
- Leave of Absence
- Separation
- Learning

Intelligent Services Subscribers

- Calibration
- Home Page
- Learning
- Onboarding
- Performance and Goals
- Compensation Management
- Recruiting
- Succession and Development
- SAP Jam
- SAP and Third Party Integration

Figure 1.65: Events and Subscribers

Figure 1.65 shows a summary of the system that can trigger an event notification and the components that can subscribe to those notifications and then perform a resulting action. Based on the way SuccessFactors has set up the events, they can trigger both internal and external notifications. The internal notifications are sent to internal subscribers, meaning another module or process within the SAP SuccessFactors HCM suite. External notifications are sent to external systems, typically integration platforms such as Boomi or HCI. The internal or external systems that receive these notifications are called subscribers.

The internal subscribers are predelivered. SAP SuccessFactors need to predefine the activity that should occur when an internal subscriber receives an event notification. In the diagram below I have a list of the internal systems that are setup to subscribe to receive an alert whenever there is a new hire.

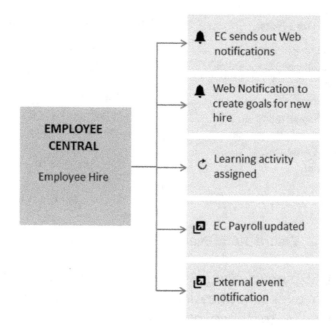

Figure 1.66: Event Notification Triggered by the New Hire Event

In the case of the first two internal subscribers, "notification of job change" and "notification to create goals for new hire," the subscribing system will generate a notification to the manager to inform them of the change. A screenshot showing some sample notifications is displayed below.

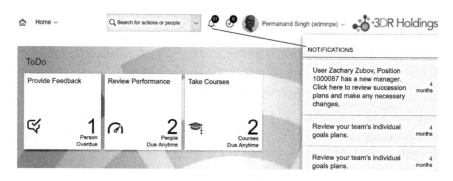

Figure 1.67: End-User Notification

If the specific internal system update that you would like to trigger is not currently available, you can still attempt to make this change using an integration platform. To use an integration platform would require the use of the external notifications. These are notifications that are published to external systems.

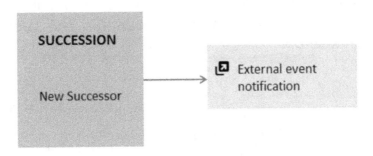

Figure 1.68: External Event Notification

The above notification from the new successor change shows an event that is only generating an external notification. In this case, this external notification can be used to trigger a Boomi process that will register the new successor for a course in the LMS. Another example can be whenever there is a leave of absence, you can use the Boomi or HCI integration platform to send an update to your benefit providers to indicate there is a change in benefit eligibility.

The use of external notifications is not limited to external actions. You can use the same integration platform to make changes back into SAP SuccessFactors. So, for example whenever there is a new hire, you can use the external notification to execute a process in Boomi that will add a specific development goal to the Career and Development Planning module for the new hire.

The setup for intelligent services varies by module and whether an internal subscriber or external system will be set up to receive the notification and trigger a process. Many of the internal subscribers can be easily setup in Admin Tools. Figure 1.69 shows a screenshot from within the event center in Admin Tools where you can enable or disable intelligent services for a particular subscriber by toggling the On/Off switch. Some of the internal subscribers such as requisition updates can require additional setup.

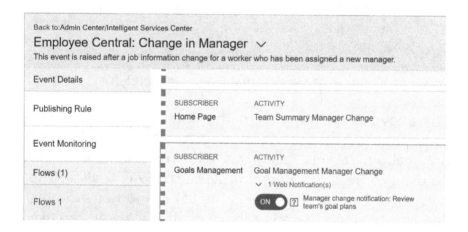

Figure 1.69: Enabling Event Notification

Setting up external subscribers is more complicated since these will require a notification to an external system. There is a separate external event notification panel that needs to be set up. This has to be done in conjunction with the setup of the integration platform or external system that will receive the notification. If you are using the Boomi integration platform, there is a web listener component that needs to be enabled to receive the event notification from SAP SuccessFactors. The screenshot below in Figure 1.70 is where you can set up the external event notification in the system.

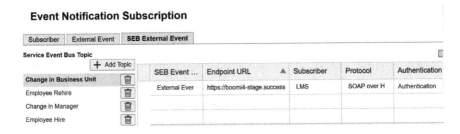

Figure 1.70: Enabling External Event Notification

Regardless if it is an internal or external event notification, you will want to track events. This is where the Intelligent Services Center dashboard is very helpful. It keeps track of the subscribing system and also keeps a running count for each event type. Figure 1.71 has a screenshot of this dashboard.

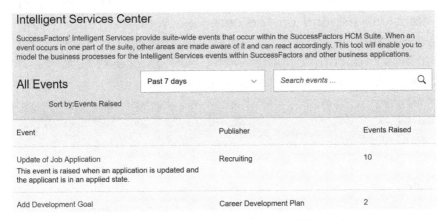

Figure 1.71: Event Center Dashboard

Intelligent Services is a great feature with a lot of potential to address unique requirements client may have. It provides additional flexibility and tools to extend the system capabilities. As you work with your consultants to design the system, do consider using this feature if the built-in data synchronization capability is insufficient or you need a specific activity to occur based on specific trigger events.

Reporting

Reporting is a critical part for any system implementation, and it serves multiple purposes. Some reports support day-to-day operational processes: open position reports, transaction reports, employee listings. Other reports provide insight into the health and efficiency of HR processes: transaction activity reports, transfer reports, promotion reports. Some can also be used to gauge the success of the overall business. This can be done by comparing current metrics with baseline values captured before the changes.

SAP SuccessFactors provides several reporting tools, most of which are built into the system. The one you select will depend on the complexity of the report, and where and how it will be used. In the figure below, the various reporting options within SAP SuccessFactors are categorized into four distinct groups.

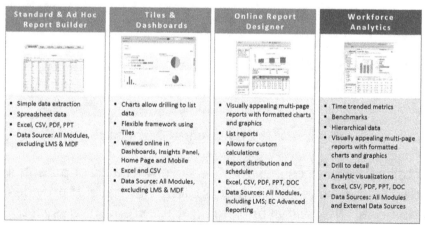

Figure 1.72: High-level Comparison of the Various Reporting Options

If you need a quick view of data stored within the system or a simple list, then the Ad Hoc reporting tool is the best fit. It is intuitive; a client system administrator can quickly learn how to use it during the system implementation so they can support the basic reporting needs of the organization once the system is live. Here is a screen shot of the Ad Hoc Reporting tool.

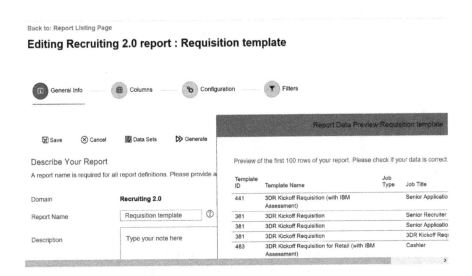

Figure 1.73: Ad Hoc Reporting Tool

Online dashboards and tiles can provide regular insights into workforce-related activity. Dashboards will vary by organization, business need, and industry, so there is no "one-size-fits-all" solution. Instead, each organization must determine the measures and indicators that are important to their business.

Dashboards are primarily created using the YouCalc tool. This is a full feature reporting tool with its own executable program so you can develop reports outside of SAP SuccessFactors. The learning curve to become familiar with YouCalc is longer than Ad Hoc or Online Report Designer, so the creation and modification of delivered dashboards and portlets is normally done by a reporting consultant who is familiar with the YouCalc reporting tool. Below is a sample of a recruiting dashboard.

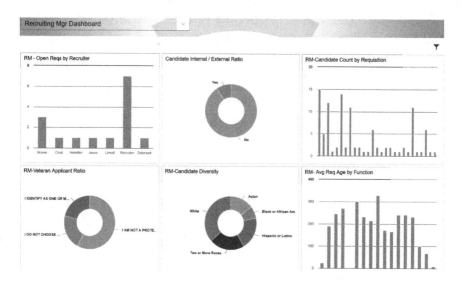

Figure 1.74: Sample Dashboard

Online Report Designer is the primary tool used to build the more complex reports that include graphics and complex computations. You can also use the BIRT (Visual Publisher tool) to generate these reports. Both are supported by SAP SuccessFactors, but I recommend that an end user or the client reporting specialist use ORD as it is more intuitive, easy to use, and there isn't a huge learning curve.

Workforce Analytics is another module that is specifically designed to identify and predict trends, establish baselines, and look at the overall strength of the organization. These reports produced by WFA typically require data that is stored outside of the SAP SuccessFactors system, so the tool needs to access other data sources.

Chapter 2

Business Objectives for the Implementation

Now that I have covered what is being implemented, this section will help define why the implementation is being done. For the remainder of this book, I will cover in detail how to complete the implementation. For many implementations, the why aspect is glossed over, but if you cannot communicate why the implementation is being done, then it will be difficult to share the importance of the project with others and define what the end state will look like.

There can be multiple reasons why an organization is implementing an HCM suite. Documenting and understanding those reasons are critical since they can provide guidance when architecting the system, defining the implementation approach, and making decisions about system design. This information can also be used to highlight the importance of the project and the system when communicating to employees.

For most clients, a full HCM suite implementation is a strategic implementation meant to support organizational business objectives. For others, it is a more tactical play when the current system can no longer adequately support HR operational needs. There are also clients who will fall somewhere in between these points. Regardless, understanding the objectives and what the end state will look like improves the odds of a successful implementation.

I will cover the business objectives for a strategic implementation first and then address the tactical implementation. The approach used to define business objectives is different based on the rationale for the implementation. For a strategic implementation, the focus is on aligning the system so that it supports HR and organizational strategy. This requires that the organization actually has well-defined strategies and objectives. This can be a very conceptual subject, so here are definitions for strategy and objective and the linkage between the two:

Strategy

Strategy is the longer term roadmap to achieve the organizational vision. It describes how an organization plans to create sustained value for its shareholders, stakeholders, customers, and employees.

HCM Strategy

Human capital management strategy is the roadmap used to describe the human capital policies, programs, practices, and initiatives to support the organization's strategic business plan.

Objectives

Objectives are the specific means to achieve the end state defined in the organizational strategy roadmap. They are concise statements similar to SMART goals (time bound and measurable).

HR Objectives

HR objectives are the specific means to achieve the end state defined in the HCM strategy roadmap. They are also concise statements describing a specific HR initiative and are similar to SMART goals (time bound and measurable)

How are strategy and objectives linked? Figure 2.1 shows a diagram representing this flow. The organizational strategy will drive both the HR strategy and the organizational objectives, which are specific and measurable objectives assigned to each business unit.

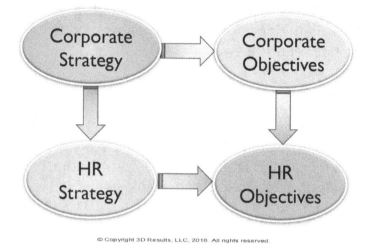

Figure 2.1: Relationship between Strategy and Objectives

For an actual example, see Figure 2.2. The organizational strategy is "Become the number one home furnishing retailer in the US by focusing on the upscale home market." This has an influence on the HR strategy, which is to become the primary employer in this sector while also improving retention through benefits and development programs. To grow the business, one of the organizational objectives is to open ten stores per quarter. The HR objectives to support both the store opening and the HR strategy include implementing a new recruiting system and reducing the time to hire. This is a simplified example meant to show how the HR objectives are supporting the organizational and HR strategies.

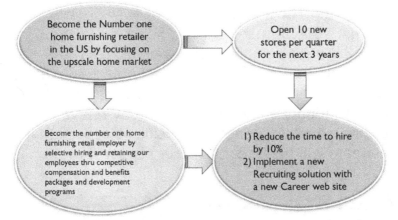

Figure 2.2: Sample Strategy and Objectives

Once specific people objectives are defined, the next step is to determine what systems and modules are needed to support them. This requires bridging strategy with the concrete specifics of timeline, scope, resources, and costs. Using a system strategy map is one way to link the two areas.

Strategy Map

One benefit of using a system strategy map is that it provides a structured approach to ensure that the talent management suite supports organizational objectives. Another benefit is that the output from the system strategy map helps to define and prioritize the various modules and features that will be implemented. There are some key activities and stages to work through when developing a system strategy map. These are:

- aligning talent management initiatives with organizational objectives;

- grouping HCM initiatives by HR function; and

- prioritizing based on need and dependencies.

To align the HR initiatives with the organizational objectives, you can use a diagram similar to the one in Figure 2.3. This is sample information from a retail client. The company has four organizational objectives that are cascaded across the company. You can see that three of the organizational objectives are directly relevant to the stores group and will generate specific objectives that the stores team will be responsible for supporting.

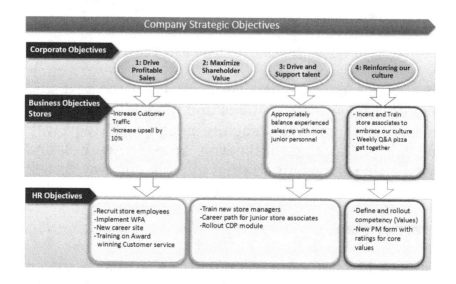

Figure 2.3: HR Deliverables Mapping

The store objectives will have a people component that is captured in the HR objectives section. There may not be a people component to each store objective or organizational objective. When building out the system strategy map, you can choose to exclude those entries or gray them out so you are focusing on the people deliverables. The same information should be captured for all business objectives that have a dependent people component so that a comprehensive talent management system landscape can emerge.

The next step is to represent the grouping of HR initiatives by HR function. Refer to the diagrams in Figure 2.4 and Figure 2.5. The first diagram shows all the HR initiatives that are supporting the organizational objective "Drive and Support Talent." A similar representation should be created for each organizational objective. It's likely that several HR projects will support multiple organizational objectives. In this case, by associating those HR initiatives with all the objectives that it is supporting, it will be easier to identify the importance of that initiative. The more times a specific initiative is listed, the more critical it is to include in the final system design.

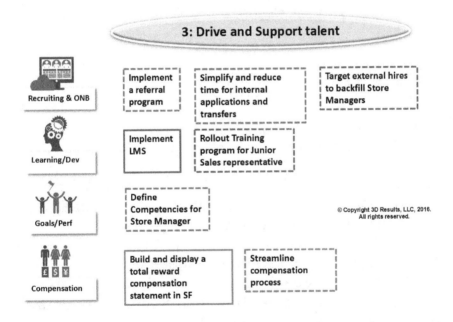

Figure 2.4: HR Initiatives by Organizational Objectives

Figure 2.5 looks at the information from the perspective of a specific HR functional area, in this case showing all the initiatives being considered for the learning and development function, grouped by organizational objectives.

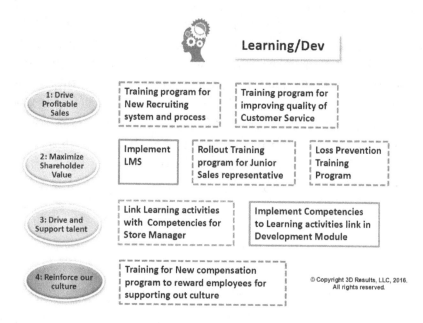

Figure 2.5: Learning and Development
Initiatives by Organizational Objectives

Similar information from all HR functions will form a comprehensive picture of all proposed HR initiatives across the different organizational objectives.

Once you can illustrate the alignment of your HR initiatives with organizational objectives and group them by functional area, the next step is to prioritize them based on need and dependencies. Completing this exercise requires more information about implementation constraints. These are tactical items that can affect priorities, such as HR focal calendar periods, resource constraints, and service contracts about to expire. These are not strategic items, but before planning out a large implementation, it is important to consider them.

A summary of a sample implementation approach is shown in Figure 2.6. The substantiating information that was gleaned from the strategy maps and the implementation constraints should also be provided to the implementation team. This way they will understand the basis for the implementation approach and can articulate how the various HR initiatives are supporting the organizational objectives.

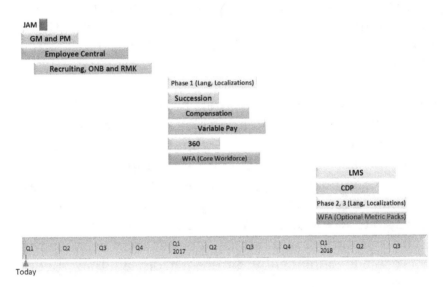

Figure 2.6: Implementation Approach

At a more tactical level, this information will be used in multiple facets of the implementation, from design to testing, training, change management, and communications. For example, an implementation team is undecided if they should include and rate competencies in the performance management form or include a development plan section instead. If the primary reason for implementing the Performance Management module was to support the "Reinforce our Culture" business objectives, this becomes a much easier decision to make. Including and rating the competencies will help reinforce culture more than development goals. Once employees know they will be evaluated on the "Reinforce our Culture" competencies, they are more likely to display the appropriate behaviors.

Guiding Principles

No matter whether the implementation is more tactical or directly linked to business objectives or organizational strategy, it is still useful to define guiding principles for the implementation. Clients who start a project without defining what they plan to accomplish will often have missed timelines and a project gone awry.

Essentially, without well-defined business objectives or clearly articulated guiding principles, the design approach tends to only focus on system capabilities and decisions around which features to use. This approach takes a long time, is confusing, and results in a more complex implementation from attempting to use as many features as possible. In this case, the configurability of the system can be a detriment and reduce user adoption because the end result is significantly different from anything users have experienced.

When implementing new features and processes, it is best to take an incremental approach based on the level of change that can be tolerated by the organization. While there will still be occasions where the project team can't make a decision around certain features, having guiding principles in place can help break the logjam. Some examples of guiding principles include:

- simple and streamlined processes to reduce effort and time spent on HR activities;

- intuitive design and ease of use to support end user adoption;

- simplicity of administration to reduce ongoing support and maintenance requirements; and

- alignment of business processes with the known architecture of the system to reduce complexity.

Figures 2.7 and 2.8 show other examples of guiding principles.

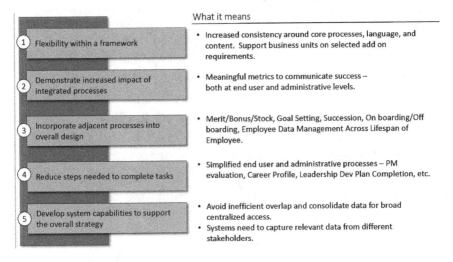

Figure 2.7: Example of Guiding Principles

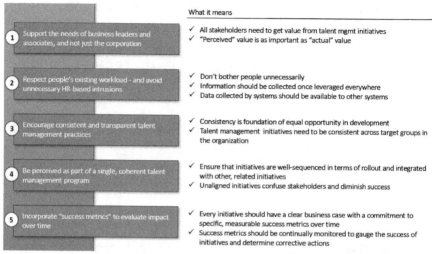

What it means

1 Support the needs of business leaders and associates, and not just the corporation
- ✓ All stakeholders need to get value from talent mgmt initiatives
- ✓ "Perceived" value is as important as "actual" value

2 Respect people's existing workload - and avoid unnecessary HR-based intrusions
- ✓ Don't bother people unnecessarily
- ✓ Information should be collected once leveraged everywhere
- ✓ Data collected by systems should be available to other systems

3 Encourage consistent and transparent talent management practices
- ✓ Consistency is foundation of equal opportunity in development
- ✓ Talent management initiatives need to be consistent across target groups in the organization

4 Be perceived as part of a single, coherent talent management program
- ✓ Ensure that initiatives are well-sequenced in terms of rollout and integrated with other, related initiatives
- ✓ Unaligned initiatives confuse stakeholders and diminish success

5 Incorporate "success metrics" to evaluate impact over time
- ✓ Every initiative should have a clear business case with a commitment to specific, measurable success metrics over time
- ✓ Success metrics should be continually monitored to gauge the success of initiatives and determine corrective actions

Figure 2.8: Another Example of Guiding Principles

The best approach is generally to define the business objectives for the system and combine them with guiding principles to define the end state and guide the system design. If it is a tactical implementation and there are no well-defined business objectives, then at a minimum establish some guiding principles that the project team can use when designing the system.

Chapter 3

Selecting the Right Implementation Partner

Other than the selection of the system, the next most important decision is which implementation partner to use. The success of the project is highly dependent on who does the implementation, and you have many choices of qualified SAP partners. In this chapter I will cover what I believe to be the key decisions and criteria to consider when selecting an implementation partner.

The Decision Matrix

To make the partner selection a structured process, I highly recommend that you use a decision matrix. Figure 3.1 shows a simplified decision matrix. When creating your own decision matrix, you would specify the selection criterion, rating scale, which vendors will be evaluated, and the members of the internal evaluation team. In this example, all the selection criteria are equally weighted.

Not shown in Figure 3.1 are the individual scores of the evaluation team members. Each member of the selection team has a more detailed area to indicate their overall ratings and pros and cons for each partner. The individual team member ratings are then summarized and included in the overall score, which is displayed in this summary view.

Criterion	Partner 1	Partner 2	Partner 3	Partner 4
Cost	1	5	2	4
Experience	3	2	2	4
Full Service Capabilities	4	5	3	5
Stability	2	2	4	2
Partnership	5	4	4	1
Culture and Fit	4	2	3	3
Post Go Live Support Services	2	4	1	4
References	5	4	3	3
Total	**26**	**28**	**22**	**26**

Figure 3.1: Simple Decision Matrix

Some organizations will choose to assign a weighted value to different evaluation criterion. It makes the process more complicated, but it is also more accurate. If references are more important than cost, then a higher weight should be assigned to references. The rating for each criterion then becomes the average rating from the selection team multiplied by the weight assigned to that criteria. Figure 3.2 has a modified version of a decision matrix with weights incorporated.

Criterion	Weights	Partner 1		Partner 2		Partner 3	
		Rating	Total	Rating	Total	Rating	Total
Cost	3	5	8	2	6	4	12
Experience	3	2	5	2	6	4	12
Full Service Capabilities	4	5	9	3	12	5	20
Stability	2	2	4	4	8	2	4
Partnership	4	4	8	4	16	1	4
Culture and Fit	4	2	6	3	12	3	12
Post Go Live Support Services	2	4	6	1	2	4	8
References	5	4	9	3	15	3	15
Total			55		77		87

Figure 3.2: Decision Matrix with Weights Incorporated

Cost

Cost is listed first as this is usually one of the key decision factors when selecting vendors. However, for

HR and cloud implementations, I recommend that cost be assigned a lower weight than experience and some of the other criterion. There are several reasons for this recommendation.

The first is that the system will be used by all employees for typical HR activities such as hiring, promotion, off-boarding, annual reviews, training, salary planning, bonuses, and so forth. Saving money on the overall implementation by selecting a lower cost partner may seem like a good idea, but if the vendor is not able to deliver a usable system, there will be a negative impact on user rollout and adoption. The loss of efficiency and benefit from the system could be greater over the long term than the savings of a lower-cost vendor.

If your system deployment is focused on supporting your organizational objectives, then you may be at even greater risk if the partner is not able to deliver. More strategic implementations require a different skill set and approach—these projects are not just about converting paper-based forms or automating a manual process. These strategic projects require a partner that understands how the system will support the organizational and business objectives. The impact of a poor implementation will be more substantial than any savings gained by using a lower-cost partner.

When considering cost, make sure your comparisons are equivalent, or apples to apples. Also, verify the type of go-live support that is provided and how long the partner will support you after the system is live. This is a critical time period for the implementation and you do not want there to be any problems while support is transitioned from the partner to SAP. I will cover support transition in more detail in Chapter 6.

Finally, even if you receive a fixed bid, ask the partners about the assumptions they used when they structured the bid. It could be that one partner has done a better job of scoping the work than another and hence has a more accurate cost.

Experience

With this important selection criterion, I believe there are two aspects of experience that should be considered—overall experience implementing cloud-based SaaS systems and the experience of the individual consultants who will actually be involved in the implementation. This is more relevant to the larger partner firms where there can be sizeable differences in the skills and experience of their consultants.

The importance of prior SAP SuccessFactors experience shouldn't be underestimated. Implementing a SaaS solution is not the same as an on premise implementation or custom development. The partner has to work within the confines of the configuration capabilities to deliver a solution that will support your organization's unique requirements and processes. While SAP SuccessFactors offers significant configuration options, not every requirement will be supported. An experienced partner can craft solutions to work around any of these limitations or alternatively redesign your process to use supported features to help you accomplish your objective.

Depending on their role (refer to Figure 4.1 in the project team role section), a consultant's skills and experience will vary. Some parts of the system are more mature than others, and you can expect to get more skilled consultants in some areas. Regardless, it is important that consultants have appropriate SAP SuccessFactors certifications and have successfully completed implementations for other clients.

In general, when I am interviewing candidates for an SAP SuccessFactors consultant position, I look at the following five areas:

1. Project management skills: even for non-project management positions

2. Communication skills: they will need to communicate with different roles on the client team, their colleagues on the partner team, and third-party vendor personnel

3. Technical skills: SAP SuccessFactors system expertise, general or specific technical skills such as Boomi or HCI based on their role

4. HR domain expertise: succession, compensation, recruiting, etc.

5. Team player: will the person fit in and support the other team members

I recommend that you use similar criterion when reviewing the consultants who will work on your implementation. Ask the partner to provide consultant bios beforehand, and have a conversation with the consultant before they are officially assigned to your project so you can address any concerns.

Full Service Capability

By full service capability, I mean partners that can provide end-to-end services for the implementation. The partner ideally should have consultants capable of implementing all modules being considered along with any impacted peripheral systems and technology platforms that will come into play. For example, there can be an integration platform used to build out integrations to other systems, the back end HRIS could be SAP or PeopleSoft, and these may need to be updated.

Within the individual modules, there are some configuration decisions that can impact other modules. For example, even if you are only implementing PM or GM, it is important to understand how information from these modules is used in Succession or Compensation so there is no adverse impact on these modules in a later phase.

When reviewing the partner's capabilities, also look at nontechnical areas where you may need the partner to supplement your resources. These can include project and program management, change management, training and documentation, and post go-live support. Some clients will create their own change management, training, and communication content, while others will use a third-party vendor or request that the partner provide these resources. Even if you are uncertain whether internal personnel will be responsible for these activities, it is still a good idea to confirm that the partner can provide support in these areas.

Depending on the scale and complexity of the implementation, also review the partner's reporting and solution architecture service offerings. These are often overlooked but critical for large, global implementations that have multiple phases. Try to find a vendor that you can continue to work with as you roll out future phases.

If a single partner cannot provide coverage in all areas of the implementation, then you may need to use multiple partners. If this is the case, then one partner should be the primary contractor and manage other vendors as subcontractors. The project ownership resides with a single partner, thereby avoiding any ambiguity or passing the buck.

Stability

You also want to select an implementation partner with a record of longevity and stability. This provides assurance that your partner will be available beyond the first phase for subsequent phases and for post go-live support. Changing vendors midway through a project increases the chances of a failed implementation, or at a minimum increases costs. When looking at stability, there are several key factors to consider:

- years in existence as a company

- years implementing SAP SuccessFactors

- financial strength

- consultant and leadership turnover rate

You'll want to understand how long the company been in business as well as the length of time they have been implementing SAP SuccessFactors. Is this a new line of service for them? If so, what is their commitment to their service offerings in this area and to developing their consultants?

To evaluate financial strength, your procurement team will want to review the company balance sheet and cash flow. In the case of a private company, you can request a call with a financial representative from the partner to get some insights into their financial stability.

Short tenure for a partner's employees should raise a red flag. You don't want to lose a key consultant midway through the project. You can request an HR-related conversation about how their methods of employee development and retention.

Partnerships

Your implementation partner may have its own partnerships with third-party vendors. For example, if you use the Dell Boomi integration platform, you would want your implementation partner to be a Boomi partner and Boomi certified. The partner agreement would ideally include implementation and support.

As an additional benefit, these partnered organizations would have worked together before and have built a collaborative relationship. They're also not learning on the job to make the two systems work together—they've done it already. I believe that partnership with other vendors demonstrates further commitment to investment in the HCM space.

When using a third-party product or service that is being resold or implemented by your implementation partner, confirm who is responsible for the support once you are live. If support is provided by the third-party vendor, then confirm at what point in the project this transition will occur. Ideally, try to have some overlapping support coverage for a few weeks so the new team can become familiar with your configuration and personnel before they become the primary point of support.

Culture and Fit

You will be working very closely with your implementation partner's consultants for an extended period of time—from six months to over a year, depending on the complexity and scale of the project. Even though most SaaS implementation work is done virtually, you will still have daily communication, meetings, and onsite reviews, so there will be significant interaction between your employees and the partner's consultants. It is important that you are comfortable with their approach, culture, and values.

Some partner firms are more customer-centric than others, with a higher willingness to engage in collaboration and problem-solving. Assess how well the partner communicates their vision and values and how they empower their employees. This will help demonstrate their accountability for results.

It's also important to assess the partner's implementation approach. Are they using an agile methodology with the recommended three iteration configuration phase, or is it a more rigid design, build, test, and deploy approach? Will the partner's implementation approach be compatible with your internal project management standards? If not, will they work with you to explain or address the differences? If this is your first cloud-based system implementation, be prepared to be flexible since the implementation activities are different than those for an on-premise or in-house developed system.

Try to learn how the consultants will approach the implementation. Will they take a more tactical approach by focusing on the configuration workbooks? Or will they take a more consultative approach by focusing on the business objectives of the implementation? If you are making the changes to an existing system, the focus on configuration workbooks could be more appropriate. If this is a new implementation, focusing on how the business will be using the system would be a better approach.

You also need insight into how your own organization works. It is not just about finding the best implementation partner; it is more about finding the optimal implementation partner for your organization.

Post Go-Live Support Services

In this section, I'll look at the support activities needed after the system has been implemented. SAP SuccessFactors offers two levels of support: SAP Enterprise Support and SAP Preferred Care. The latter will provide you with more coverage but may not cover all aspects of your system. The support transition period is critical to the project success, but is often an uncomfortable time since it involves changing your first line of support as more employees begin to use the system.

Until the production sign-off has been received, the partner is generally responsible for all changes and support of the system. Once the client has provided production sign-off, support activities transition from the partner to SAP SuccessFactors.

To alleviate the pains of the transition period, you may want to engage your partner for extended support during go-live. This can be done by having your consultants available for several weeks after the launch or by having them provide dedicated administrative support for your internal team.

While SAP SuccessFactors provides support for the system and the Boomi platform, making modifications to existing configurations, custom integrations, data issues, and reporting are not covered. This is why it is critical to have the consultants who were involved in the implementation available to assist if there is an issue. They can troubleshoot and resolve an issue a lot more quickly than the SAP SuccessFactors support team who won't be as familiar with your configuration. In any case, if you require modifications to existing configurations, custom integrations, data issues, and reporting, SAP SuccessFactors will request that you contact your implementation partner to resolve these issues. You'll want to discuss the partner's capabilities to provide post go-live support as part of the selection process.

References

References are a key part of the vendor selection process. A well-conducted reference check can give valuable insights into a partner's implementation capabilities along with their culture and value. Your company may have a process and a recommended set of questions to use for conducting reference checks. If not here are some suggestions for conducting your own reference checks.

When requesting a reference, ask the partner to provide organizations similar in size and within the same industry as your own. The reference feedback will be more relevant, and you may also get some unsolicited ideas about how to make your implementation more successful based on what the reference organization has learned.

Going into the reference call, be aware that the partner will provide you with their best references, who will highlight areas where the partner has excelled. The key to a thorough reference check is to ask open-ended and hypothetical questions. Here are some sample questions. The first question is an important one in case there were several partners involved in the implementation.

- What role did the partner play in the implementation?

- Why did they choose this partner over others?

- Who were the other partners they reviewed?

- If they had to do the implementation again, what would they do differently?

- How did the partner help them to implement the system on time?

- Were there any situations where the partner took a consultative or creative approach to addressing the client's pain points?

- Were there any undesired or unplanned outcomes?

- Will they provide some example of where change orders were used?

When doing reference checks, try to use similar questions with all the references so you can compare the responses. You can use a matrix similar to the decision matrix that was covered earlier in this section. One additional recommendation is to group the questions into categories such as general, implementation, technical, personnel, and so on. This way the responses will be easy to review and compare with the other references.

Chapter 4

Preparing for the Implementation

As a large system comprised of multiple modules with features that will be used by all employees, SAP SuccessFactors requires significant internal effort from your organization for a successful implementation. It is critical that the right internal team members are selected to work on the implementation and that they have the available time to do so. Often, clients will have core team members remain fully responsible for their day jobs with the implementation as an additional responsibility. This may work for subject matter experts or advisors, but is not viable for core team members.

See Figure 4.1 for a list of all the various roles involved with the implementation. You will notice that many of the roles are not related to the technical design and configuration aspects of the implementation. This is where the discrepancy in the personnel requirements often occurs since clients tend to focus primarily on design and the configuration and not the other areas. Yet, it is these other activities that are often overlooked during project planning that take the most effort.

Your internal project team will need to be involved in activities such as project management, vendor management, design, integration, and testing. The activities that are not obvious include the internal change effort, content development (such as competency modelling, learning activities), process optimization, training, and communication. These activities are often not considered when personnel are allocated to the project and then core team members are tasked with multiple responsibilities and overbooked.

There are many activities and tasks in a typical SAP SuccessFactors implementation that will be included in the project plan. Your implementation partner will work with you to create the baseline project plan and to staff the project. If you don't have adequate personnel on your internal team to fill a role, your implementation partner may be able to supplement your resources. Activities such as training, communications, change management, and project management are areas where the partner can provide additional assistance.

The Project Team

To assemble the right team, it is best to first look at the personnel requirements from a roles/responsibilities perspective and then, based on expertise and availability, assign individuals to the various roles. This will ensure that you have coverage for all the critical areas. Figure 4.1 shows the typical roles involved in a large or global SAP SuccessFactors implementation.

Role	Accountability
Project Sponsor	• Champion project, validate decisions and assign resources
	• Serve as an escalation point for critical decisions
Project Lead	• The business/process owner of the project, responsible for decisions, collaboration, design, resource assignments, etc.
	• Approve critical project milestones
Project Manager	• Manage/coordinate internal resources, timelines, and deliverables
	• Primary project point of contact
	• Manage and escalate open issues and risks for resolution
	• Test configuration within the system after each iteration
System Administrators	• Own and administer system upon go-live
	• Test configuration within the system after each iteration
	• Attend system admin and reporting training classes
Design Team (Functional & Business Reps/SMEs)	• Approve recommended process and configuration decisions
	• Test configuration within the system after each iteration
	• Test the system during User Acceptance Testing (UAT)
	• Will be process and workflow experts after implementation
	• Provide input/validate process, design, and workflow during UAT
	• Provide input that informs and guides change management plan
Technical Reps	• Coordinate data file creation and testing of FTP, SSO, employee data file load, and any imports and vendor integrations
	• Participate in all data and integration discussions

Steering Committee	• Provide input and guidance for overall direction of project
	• Monitor project progress and serve as escalation point for major project decisions and risks
	• Point of contact for key stakeholders
Advisory Council	• Provide input and feedback on business process, design, and configuration decisions
	• Serve as communication link back to business and users about the project
	• Provide input and advice on change management and training needs
Security Lead	• Work closely with the business process owners to develop security plan for managing data and role based permissions in line with company policy
Testing Lead	• Test configuration within the system after each iteration
	• Manage/coordinate SIT and UAT activities
	• Triage test issues identified from UAT
Reporting/Analytics Team	• Work closely with Core project team to determine reporting requirements, tool functionality, and resourcing for report creation
	• Test and validate reports created within the system
Change Management Lead	• Develop change management plan and ensure implementation of the plan stays on track
	• Advise project manager and process owners on change management and stakeholdering issues
Communication Lead	• Test configuration within the system after each iteration
	• Define/coordinate communication needs with Change Management Lead
Training Lead	• Test configuration within the system after each iteration
	• Define/coordinate end user training needs with Change Management Lead
Legal Representative	• Provide Input/validate content throughout engagement

Figure 4.1: Project Team Member Roles and Responsibilities

Assembling the Right Team

Now that the roles have been identified, the next step is to assemble the team and assign individuals to each role. Your implementation partner can provide some estimates on approximate time commitments for each role. Based on the time commitments, availability, and skills needed, you can round out your internal implementation team.

From experience, the teams that were most effective at managing an SAP SuccessFactors implementation as well as driving the business process decisions necessary for success had the following characteristics:

- were led by a full-time project manager with experience leading IT projects;

- included HR and IT subject matter experts, business representatives, key stakeholders, and executive sponsors who were actively engaged in the project;

- were guided by a steering or advisory committee;

- ensured that team members had clear roles and responsibilities, and that each team member had both the capability and capacity to execute those responsibilities;

- created a project team governance structure that defined how the team would work together and which team members had a voice, vote, or veto;

- ensured that the team was able to effectively and efficiently make decisions to keep the project moving forward; and

- ensured that team members used their time together effectively and productively.

The ideal project team should include representatives from IT, HR, the business, and senior leadership. The project manager should have experience in systems implementation. Ideally they should come from your project management office or from the IT department. Having dual project owners where one is from HR and the other from IT is also a good recommendation. Both HR and IT senior leaders should share sponsorship, if possible. If there is only one project sponsor, that role should be filled by the senior HR leader. HR should provide subject matter experts (SMEs) for all business processes; representatives from the business should also play an SME role where appropriate and feasible, and ideally with global representation if at all possible.

In addition, use an advisory council or steering committee (some organizations have both) made up of representatives from the business, HR business partners, key project team members, and other stakeholders to provide input into and validate business process design decisions and offer feedback on the proposed solution throughout the project. Figure 4.2 shows a sample project team structure.

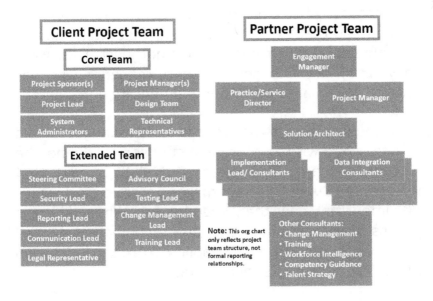

Figure 4.2: Typical Client and Implementation
Partner Project Team Structure

Once the team has been assembled, it should be documented and the appropriate adjustments made to their daily responsibilities so they can focus on the implementation. The list of roles in Figure 4.1 can then be updated with two additional columns to include the time commitments and individuals associated with each role.

Project Management

A key role for the project's success is an experienced project manager. I cannot overestimate the importance of this role. This individual will have to drive change, accountability, and decision-making during the implementation. In a multiphase project, the project manager can also help with the transition and knowledge sharing from one phase to the next since there will be personnel changes between phases.

If your company has a project management office (PMO), I recommend that they provide the project manager for the implementation. If a PMO does not exist, then work with your IT team to get a project manager who has managed other large system implementation projects.

One of the first things that your project manager should do is to work with the project sponsor to define the governance team. The integrated data model that underlies SAP SuccessFactors makes program governance more important now than ever before. When different business functions, processes, and technologies existed in silos, each group had the ability to manage its own data, software updates, and changes.

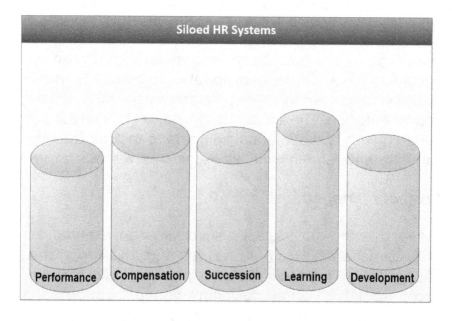

Figure 4.3: Siloed HR Ownerships

Figure 4.3 shows a visual representation of a siloed HR system. Each HR function has its own system and management. With the implementation of SAP

SuccessFactors cloud-based HCM suite, all these systems will be integrated into one solution. This presents two significant changes for some organizations. First, they are now moving from an on-premise system to a cloud-based solution. Even if you have used a governance model in the past, there are some additional considerations for cloud-based systems. These differences are highlighted in Figure 4.4

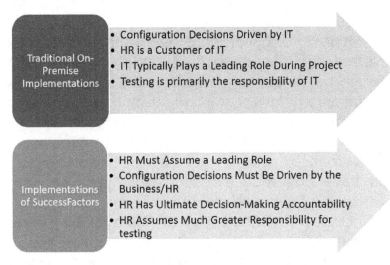

Traditional On-Premise Implementations
- Configuration Decisions Driven by IT
- HR is a Customer of IT
- IT Typically Plays a Leading Role During Project
- Testing is primarily the responsibility of IT

Implementations of SuccessFactors
- HR Must Assume a Leading Role
- Configuration Decisions Must Be Driven by the Business/HR
- HR Has Ultimate Decision-Making Accountability
- HR Assumes Much Greater Responsibility for testing

Figure 4.4: Governance Differences in On-Premise Versus Cloud

The second significant change is the move from siloed HR systems to a single integrated HCM suite. Now that the system is shared across the organization, teams must work collaboratively to design it and make decisions that are beneficial to the whole organization and not just their particular function. These governance decisions can extend into data security, access, and business process changes. Initially, a governance structure will be more time-consuming since it may be a change to how HR projects are managed in your company, but this is actually an additional benefit of implementing an integrated HCM suite. It helps you to break down these silos and naturally extend your business processes across multiple HR functions. For example, a new hire process can extend beyond the original confines of recruiting and could impact what goals, development objectives, and learning are assigned to the new hire based on information gathered through the screening and onboarding process.

Once the project manager has been identified, then defining the governance team is the next step. Who should be on the governance team? The first decision is to determine if the governance team will be responsible for system decisions both during the implementation and beyond go-live. Many organizations create an initial governance body for the implementation and evolve the team to a governance committee once the system is fully implemented. Figure 4.5 shows a typical composition of a governance team. In general, because this team should have governance responsibility for the entire SAP SuccessFactors platform in your organization, it will ideally include representatives from HR and IT as well as the various lines of business. Business representatives will also be in the best position to share requested system changes and seek out opportunities for improvement both within the business process and the tool.

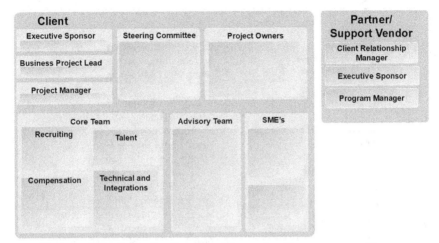

Figure 4.5: Sample Composition of a Governance Team

Steering Committee

One group mentioned in the overall governance model is the steering committee, so I'll look at the function and value that this group can provide to the overall project success and why it is important to secure its active participation.

The primary objective of the steering committee is to provide strategic direction and major issue resolution around business alignment, project design, and implementation. The committee members are also jointly responsible with the project manager for project results. Figure 4.6 shows the typical process for how unresolved decisions are pushed to the steering committee.

Core Team
- Daily interaction with the implementation partner
- First level of decision making

SMEs & Advisory Team
- Depending on the project phase, they will have weekly or more frequent interaction with Core Team
- Second level of decision making authority

Sponsors & Steering Committee
- Monthly meeting and impromptu escalation meeting if needed
- Final level of decision making authority.

Figure 4.6: The Steering Committee

The steering committee members are senior personnel in the company and likely will have limited availability; hence it is important to maximize their interactions. You can also help to get their commitment by defining what the benefits of their participation, time requirements, and responsibilities.

When discussing the benefits, it is best to focus on the impact the project will have on organizational objectives and/or their business unit. The strategy maps and business drivers discussed in Chapter 2 can help drive this conversation. If the project is going to impact their specific business unit, remember to ask them to commit their personnel to the project team. Your user adoption will be greater if representatives from the business are included in the project team and actively involved in design, testing, training, and communications.

You will want to minimize the time commitments of steering committee members. Two to three hours per month is realistic—an hour for the monthly steering committee meeting and additional time for impromptu requests, e-mails, and any escalations. Here are some typical responsibilities for the steering committee:

- accountable for successful delivery of project

- provide oversight and serve as escalation point for issues and decisions

- approve project scope changes

- serve as organizational sponsors and project champions—communications/change management liaison to other leaders

- provide strategic vision

- decision makers for resourcing and project prioritization

- ultimate project decision makers as a committee (single body)

There are multiple ways the steering committee can help achieve a successful project completion. In addition to their main responsibilities, steering committee members may be able to help get the right personnel on your team, approve funding and scope changes, communicate the benefits of the project throughout the organization, and enforce user adoption.

Work Streams

In any large project implementation, having a single, large project team is inefficient. Creating and using subteams makes better use of team members' participation and also alleviates confusion and delays. Some HR functional areas are very different than others. It is almost like moving to a different job entirely if you ask someone in Learning to work in Recruiting or Compensation. By having separate subteams focus on their particular domain, it reduces this confusion. Smaller groups also facilitate scheduling meetings and gain team consensus on critical decisions.

When using subteams, begin by identifying the needed work streams. This way, each work stream can have its own team. The division could be based on the SAP SuccessFactors module, by department, by business process, or some combination thereof. A lot will depend on your organizational structure along with the size, complexity, and length of the implementation. The larger and more complex the implementation, the more granular the work streams will be. For a phased project with less complexity, the same team can be used across the difference phases where appropriate.

Here is a typical subteam setup for a full suite global implementation. Note that since the implementation will be done in phases, not all the teams will be active simultaneously, though their lives may extend beyond the implementation time period since they can also be focused on pre-implementation process standardization and post-implementation support activities.

- recruiting (RMK, RCM, ONB)

- compensation (Comp and VP)

- performance and Development (PM, GM, 360, CDP, Succession)

- learning

- technical (Data, Integration, SSO)

- platform (Competencies, Job Profile Builder)

- reporting

- change Management

- regional Subteams (based on geographic location)

Change Management and Communication

Change management is often overlooked in a technical system implementation. For a small implementation or one where the system is only used by a single department, the impact of not implementing a change management program may not be as pronounced. For a large system implementation and, in particular, one that will be used by all employees, it is important to prepare the organization to accept the changes. If this is not done, the overall success of the project will be in jeopardy.

Change management brings a structured approach to comparing your current state with the future state to identify what is changing, why the change is needed, and to prepare the organization so they will more readily embrace this change. In short, the change management team will help you maximize the adoption of SAP SuccessFactors by minimizing the users' natural resistance to change.

If your organization has a change management department, inform them of this project at the earliest opportunity so they can assist with the rollout. If you do not have a change management team, you can ask your implementation partner to assist with your change management program or have them recommended another company that specializes in this area.

To implement effective change management programs, first identify the team. As mentioned before, this can be from your internal team, or it could be from the partner or another consulting company. Then define the roles and responsibilities for the team and its members. Define the vision and set expectations as to what the change management team will accomplish. Typically, the objectives of the change management team are to conduct a thorough, real-world analysis of the impact of your SAP SuccessFactors implementation and develop the governance, communications, training, support, and measurement plans to support the rollout.

To do this, they will address the following questions:

- What aspects of your organization's culture might impede or help a change?

- What is actually changing to process and policy?

- How big is the gap between current and future state?

- Which stakeholders will be most impacted?

- How can we best facilitate their transition?

- What are the risks that could derail this implementation?

Based on the responses to these questions, the team will build out the change management plan. This plan can either be included in the master project plan or exist independently. The activities and tasks in the plan will typically be grouped in into different categories like those listed below. This is graphically represented in Figure 4.7

- communications

- training

- post go-live support

- measurement

- governance and feedback loop

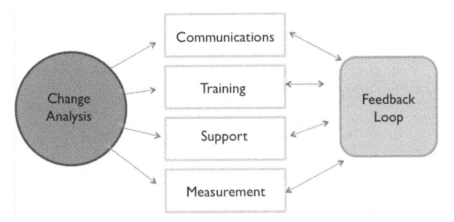

Figure 4.7: Key Areas in the Change Management Program

The change management team will develop and distribute the content, communications, and risk mitigation plans to maximize adoption of your SAP SuccessFactors implementation and help ensure adoption of the new system and processes.

Chapter 5

The Implementation

By now, you should have had conversations with your implementation partner about the project plan, identified the internal project team, and be ready to start the implementation. The next major milestone is the kickoff meeting. Your partner will work with you on who should be present for this meeting and what the agenda will cover. The type of kick off meeting (big bang with all the project participants or a low key affair that is confined to a few key participants) will depend on the implementation approach.

If it is a small or tactical implementation, then it is likely that the project will start out with some preplanning to determine the key dates, resource planning, and the kickoff meeting with all the team members. This will be followed by the prepare and the realize stages (see Figure 5.1 below, which uses the SAP Launch Methodology). Following these stages will be the verify stage for testing and then the go-live or launch stage.

On the other hand, if it is a larger or global implementation, more planning, process analysis, and gap analysis are done up-front in the prepare stage. A larger implementation requires more defined objectives around what you are looking to accomplish and a more detailed roadmap to achieve the desired end state. It is lot easier to lose your way when implementing a full HCM suite over three years as compared to a two module implementation in a six-month time period.

Your implementation partner may embellish this methodology or have a hybrid approach to implementing SAP SuccessFactors to address some of the more comprehensive planning activities that are required for a large implementation. This may include an expanded prepare stage to document current state processes and then design the future state. It can also include detailed, solution-oriented activities such as designing the system architecture roadmap and defining the implementation approach. For multiphase projects, it is more critical to have these types of high-level design documentation so you will know what is included in each phase and can plan accordingly.

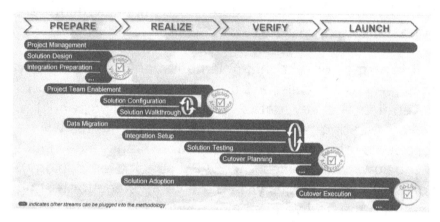

Figure 5.1: SAP Launch Methodology with
Key Stages in the Implementation

Each stage of the implementation approach will end with a key milestone and a quality gate check, where you will be asked to review the completed work and sign off on it before moving to the next stage. Here are the key sign-offs needed for each stage based on the SAP launch methodology. As mentioned before, your implementation partner may have a different methodology with different quality gate checks.

Prepare: This stage focuses on the planning and design of the system. For a large global implementation, there can be an initial approval for the overall implementation approach and architecture roadmap. For a smaller implementation, you may jump directly into the configuration design. Regardless, the signoffs in this stage are focused on the system design.

Realize: The deliverables and signoffs in this stage are focused on the actual system configuration. You will confirm that the system is configured based on the agreed upon designs that were approved in the first stage.

Verify: This is essentially two separate signoffs that are often combined into one. The first confirms that you have completed full user acceptance testing, and the second is to indicate that you are ready to proceed with the cutover to the production environment.

Launch: For this signoff you are asked to confirm that the implementation project is complete. All modules are correctly configured in the production instance and are ready to launch to your users. The service delivery team will then start to transition support over to the SAP support team. I will cover the support transition in more detail in Chapter 6.

I will not cover all of the implementation stages in detail. This will be done by your implementation partner based on their own implementation approach. Instead I will provide more details in a few select areas that are different than a typical custom project or ERP implementation, namely the design, three iteration approach, testing, and production validation activities.

Design

System design is done differently in an SAP SuccessFactors implementation versus a custom developed system or ERP implementation. The design process is much simpler because you don't start with a blank page or have an infinite set of design options from which to choose. With that being said, it will still take some time for you to work with your consultant to review and decide on the optimal configuration for your organization.

To simplify the design process, most partners will provide some baseline or recommended starting configuration. This will be reviewed with your team, and the proposed changes will be documented into the configuration workbook, which is what is used to capture all the design information. The primary design will be done in the prepare stage. But since the configuration is done in the realize stage, there is an iterative three-step process (this is detailed in the next chapter) that will provide the option to still make tweaks to the design after each configuration iteration.

There is a separate workbook for each module. Once all the configurations have been captured for the initial iteration, you will be asked to review and approve the design. The consultant will take some time to configure the system based on the design options defined in the configuration workbook. Figure 5.2 and 5.3 show some screenshots of the recruiting configuration workbooks.

Figure 5.2: Introductory Page of the
Recruiting Configuration Workbook

Iteration 2 Change / Iteration 3 Change	Template Name: SF internal field	Requisition Field Labels	Standard or Custom	Type of Field (Text, Textarea, Picklist, Date, True/False, Percent)	Picklist Name (enter values on picklist TAB)	If Picklist, parent picklist	Required
	instrInformation	Requisition Information	Custom	Instruction			
	id	Requisition ID		Auto generated			
	filter2	Country		picklist	postingCountry		Yes
	filter1	Location		picklist	companyLocation	postingCountry	Yes
	division	Division		derived			
	department	Department		derived			
	location	Location		derived			
	functionalArea	Functional Area	Custom	picklist	functionalArea		Yes
	addRep	Additional Hire/Replacement	Custom		addRep		
	costCentre	Cost Centre	Custom	text			
	contractType	Employment Type	Custom	picklist	contractType		
	eeoGroup	Job Level (Required for US only)		picklist	FSLA_Status		

Figure 5.3: Requisition Fields to Be Included in the Design

Three Iteration Build

The three iteration build is a different concept than what is found in traditional custom development or ERP implementation. In these system development projects, there is only one development and one testing cycle. In an SAP SuccessFactors project, a three-step iterative approach is used to define and finalize the design and development. Each iteration is not meant to be a full design and development cycle. They are meant to build on each other.

The objective of the first cycle is to capture most of the designs so the development work can start. Iteration two and three are used to validate the design and make any tweaks or changes before the system is ready for full user acceptance testing (UAT). Figure 5.4 shows the activities in each cycle and how they flow into the next cycle. Each cycle commences with documenting and approving the updated design requirements in the configuration workbook. Then the consultant will configure the system and do a joint review with your team. Finally, your team will test the configuration and document any changes needed for the next cycle. At the end of the third cycle, and once the configuration signoff has been received, the system is ready for UAT.

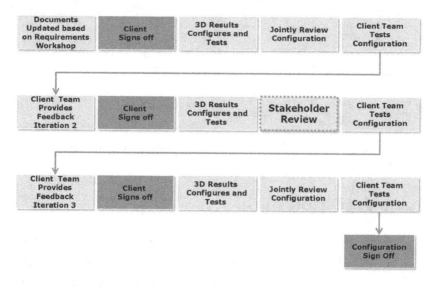

Figure 5.4: The Three-Step Iterative Development Process

For the first iteration review, the consultant will guide you through the configuration options. They will use the configuration workbook and the baseline system configuration that is loaded into your test instance to guide you through this process. This first iteration includes a small core team or work stream members who are very familiar with the requirements and processes associated with the particular module. The objective is to capture approximately 80 percent of the requirements so the consultant can work on the configurations for the first iteration.

The second iteration typically includes a larger audience. At this point, the core team has reviewed the configuration from the first iteration and made their recommendations for the changes needed for the second iteration. They will want to receive confirmation from the larger team on what they have accomplished in the first Iteration and their proposed changes for the second iteration. For the second iteration design, the goal is to meet approximately 95 percent of the requirements.

During the third iteration you will normally involve the project sponsors. You have completed testing of the second iteration and identified any remaining changes for the third iteration. Before communicating this to the consultant, you will conduct a high-level strategic review with the sponsor and the steering committee. Since this is the last iteration, you want to get approval from the steering committee before committing to the final changes. Often, clients will wait until the third iteration has been completed to review the work with the project sponsor. But if the sponsors request changes at this point, you would need an additional iteration from the partner, which can increase the project duration and cost.

I recommend ensuring that the project team is actually testing and providing feedback for each iteration and not just rubber stamping the iteration review approval. If adequate testing is not done during each review, then there will be significant changes in the third and final iteration, or worse, the requirement gaps will only be identified during the UAT, which can severely impact your go-live date. By conducting comprehensive reviews during each iteration, the project team becomes more familiar with the system and can better guide the training team to develop the content and training program.

If there are concerns about validity of the testing in each iteration, you can request a written confirmation from the core team that they have tested the iteration and it matches their design requirements. Since your project lead will have to formally approve the changes for each iteration before the consultant will commence working on them, I suggest that before that approval is given to the consultant you receive a formal acknowledgement from your internal team that they have tested and approved the changes.

Testing

The primary focus of testing is to verify that the system reflects your approved design configurations. From the SAP launch methodology (Figure 5.1) the full UAT is done in the verify stage. There are, however, multiple types of testing that occur at different stages throughout the project. Before I dive into the details, I'll look at the main types of testing done during an SAP SuccessFactors implementation. I have not included load testing, because while it is a valid part of the overall testing strategy for a system implementation, in the SaaS architecture the vendor is responsible for the scalability of their system and for providing the appropriate metrics to show how the system performs during stress testing.

If your organization has a quality assurance department or dedicated testing personnel and procedures for validating the system configuration, then you should engage with them early in the implementation to ensure they have all the information and training needed to assist with your testing phase.

Unit Testing

This is done by the implementation partner as they make changes to the specific components in the module— for example, a portion of the performance review form, workflow changes, a goal plan, or the complete performance form. The consultant will complete an initial review to ensure that the changes are what was requested by the client in the design documentation. If integrations are being developed, then unit testing will be more at the component level to ensure each part of the process is working before the full integration is assembled.

Module-Specific Testing

Since the SAP SuccessFactors HCM suite is comprised of multiple modules, it is import that each module is tested before proceeding with full UAT. The module-specific testing is initially done by the implementation consultant and then transitioned over to the client so they can complete their own test routines. Embedded elements such as goals and development plans that are in the performance review forms should also be included in the module-specific testing. At the module level, unit testing is typically done during each iteration cycle once the consultant has completed the configuration based on the approved requirements.

Parallel Testing

This type of testing involves having identical data entered into two separate systems so you can verify that the output is the same. This is typically done for any payroll and benefit system where you need to be 100 percent sure there will be no adverse impacts to an employee salary, benefits, or dependents once you switch over to the new system. To ensure the integrity of parallel testing, you need to use a separate instance and not the SAP SuccessFactors test instances. The parallel SAP SuccessFactors instance will be linked to the payroll quality assurance environment and will be compared with the existing HR or payroll system. If parallel testing is required, an additional two to three weeks should be added to the overall testing and project timeline. More robust testing scenarios should be developed that will contain the actual data that is entered into both systems. Typically, you will use a subset of the actual production data in the existing system to test out the new system. It is best to document the actual data that will be entered into both systems so you can review the output. Figure 5.5 shows a sample of the granular data needed for parallel testing.

Event	Last Name	First Name	NI Number	Address 1	Address 2	City	County/State	Country
New starter perm US	Ren	Heley	G6496969N	29 edem Roed	#02-06, edem F	Singapore		Singapore
New starter perm UK	Prece	Emely	JZ627026e	46 Westrey Welr		Westford	Essex	United Kingdom
New starter Temp	Coor	Lezzee	NB464407B	2 Porter Roed		Beton	Essex	United Kingdom
Change of address	Jones	Well	G4543303K	24 Reversede Welr		Sutton	Surrey	United Kingdom
Change of address US	Hetz	Emme	W994943B	42 Burneston St		Creenburn	Melbourne	Austrailia
Change of name	Berer	Jennefer	56595944					

Figure 5.5: Granular Testing Data for Parallel Testing

System Integration Testing

If there are integrations involved or other systems that will be affected, then system integration testing is needed to verify that all the integrations are working as expected. Integration testing is more complex because there are multiple points of failure and some of them will extend beyond the HCM suite. In addition to the testers, you will also need the integration developers, vendors, representatives of potentially affected system, and experienced users to be involved in this type of testing.

Integration testing requires you to look at other aspects of the system that are not as obvious. This is particularly true in the area of security since information is being shared across multiple systems. Items to confirm would be:

- Is the data transmission secure?

- If information will be temporary stored on a SFTP server, are the files encrypted?

- Are the SFAPI accounts permissioned correctly and access limited to specific IP addresses?

- Have all the data transformation rules been validated?

- Are downstream system actions being triggered correctly based on the integration?

The requirements and information used for parallel testing and integration testing are more comprehensive and need to be tackled differently. For integration testing, I have found it easier to track all the integrations in a separate workbook as opposed to the project plan. If you have several integrations and attempt to capture the detailed task level information in the project plan to manage these integrations, it will expand the plan by over a hundred new tasks since you will have to cover design, development, testing, production migration, and validation for each of them. If you have twenty or thirty integrations, then it becomes unmanageable to track these in a project plan.

It is easier to track the integrations in a worksheet and the use the same format to manage the testing. Figure 5.6 shows a view of the summary page of the integration worksheet.

Integrations											
Item	Vendor/System	Purpose	Direction	Main Process Name	Output File Name	Frequency	Method	Status	Next Milestone	Partner Lead	Client Lead
1	Stang	Pre-employment assessments	Bi-Directional					Design			
2	HireRight	Pre-employment background checks for MPC	Bi-Directional					Development			
3	GIS-BKG	Pre-employment background checks for Speedway	Bi-Directional					Testing			
4	Medgate	Pre-employment drug screens for MPC	Bi-Directional								
5	GIS-WOTC	Tax credit (Pending CO approval)	Bi-Directional								

Figure 5.6: Summary Page of the Integration Worksheet

Using this worksheet, you can easily identify the integrations that are in the testing phase. You can also color code the status to highlight which ones are ahead of or behind schedule. It is much easier to manage a large number of integrations using this approach instead of a project plan.

To keep track of the open testing issues, you can use a UAT issue log similar to the one shown in Figure 5.7. Instead of the module name, you can use the integration name. This allows you to use a single UAT log for the SAP SuccessFactors module and the integrations.

Issue Number	Date Reported	Identified By	Assigned To	Module/Template or Integration Name	Priority	Status	Issue Description	Comments/ Resolution

Figure 5.7: UAT Issue Log

There are other aspects of system integration testing that I will cover in the test plans and test scripts section.

Regression Testing

In a multiphase implementation, you will often need to make changes to an existing module to support new requirements from another module being implemented. For example, Employee Central could have been implemented first, followed by Recruiting in a later phase. Since Recruiting is feeding new hire information to EC, this may require some changes to the EC configuration. In this case it is always a good idea to do some regression testing in EC to make certain that neither EC nor downstream systems, such as payroll, are adversely affected.

User Acceptance Testing

The final and most comprehensive testing is user acceptance testing (UAT). The purpose of UAT is not to test the system for the first time but to complete a comprehensive testing of the system and processes. A holistic view of the system is important to successful UAT. Simply testing system functionality does not validate that business requirements are being met.

It is at the conclusion of UAT that the client will provide their testing sign off to indicate that their system is fully configured based on design specifications. This signoff also typically includes the approval needed for the implementation consultant to start migrating the configuration from the test instance into production.

There are several key considerations when planning and executing your UAT. For starters, there are some logistical questions:

- Who will be doing the testing?

- Where will they be situated (reserved conference room or in a remote office)?

- When will this be done (predefined time or flexible)?

- Which instance will be used (test or is there a staging or QA instance)?

Ideally, your testers should be comprised of representatives from all parties who have a vested interest in the system. The best testers are those that have knowledge of the current processes and policies or knowledge and understanding of the data they currently receive or expect to see. UAT of a system that involves multiple users cannot be done by a small group but should be large enough to cover all.

In addition to the logistical questions, the two other major considerations will be testing scenarios and the duration of the testing. My recommendation is that from the second iteration, there should be some members of the project team working on the testing script. Once design signoff on the third iteration has been presented to the consultant, then the test scenarios can be completed. I have covered testing scripts in more detail later in this section.

For the testing duration, more time is better. Rushing through testing guarantees that it will not be thorough. Enough time is needed to execute scripts and correct errors. To guarantee end-to-end testing, you will need multiple scripts executed by multiple users and the time required to run them. Also, the schedule must be sufficient to handle multiple rounds of testing if needed. Based on the project, multiple rounds of testing may be best to test all aspects of the system. Depending on the complexity and scale of the particular phase, a time frame of ten business days may be sufficient.

Test Plans and Test Scripts

Testing is critical to the overall success of the project, and there should be a separate set of tasks either in the overall project plan or in a separate testing plan to manage these activities. This will help ensure that the appropriate personnel are available and any prework is completed before the testing cycle begins. There should also be contingency time included to allow for troubleshooting and issue resolution.

One way to accomplished this is by setting up test runs. For example, with ten business days for testing, you can have two test runs, each lasting four days. Then, you would have two extra days for data entry, troubleshooting, and corrective action. The test runs are particularly relevant for parallel payroll testing where you want to have a successful test payroll run (without any changes or configurations) before migrating to production. By doing multiple test runs, you are separating one testing cycle from another to help ensure that you eventually have a clean test run.

I also recommend doing a pre-test routine (dry run) of the test and a pilot of the end-user training before starting the actual test. This will help determine if the training needs to be improved while you also get the testers familiar with the system.

For the pretest routine, the objective is to review and walk through the complete test process using sample test cases before the actual testing begins. This is particularly useful for system integration testing where multiple parties and vendors are involved. The pretest is an opportunity to uncover any process or integration breakdowns before having all the dedicated testers in place.

To do the pretest routine and conduct any valid and thorough testing, you need well-defined and accurate test scripts. Test scripts should include details such as testing user ID, security roles, objective of script, steps to complete the objective, expected outcomes, and actual outcomes. The scripts are best written by users who understand the system as it relates to their business requirements. They are a valuable tool for the creation of training materials, testing of the quarterly updates, and implementation of additional functionality.

Scripts should also clearly define the security role or user group for the testing. A sample test script is displayed in Figure 5.8. All testing should not be done with admin accounts, since this will be one of the last opportunities to ensure that your permission model is working as expected. In a nutshell, testing should be a true reflection of the daily system functions completed by your users.

Overall Test Result:	PASS: ☐ FAIL: ☐ PASS, WITH MINOR ITEMS TO ADDRESS: ☐		
Comments:	*Please enter any overall comments here.*		
Name and Role	George Matthews ☐ Hiring Manager ☐ Recruiter ☐ Coordinator ☐ Admin		
Mobile:	If it is a mobile device	☐ Provider Network ☐ WIFI	Specify:
Operating System/Browser:	Specify what Operating System (e.g., Windows 10), and what Browser / version (e.g., IE 11) you are using to perform this test.	Widows 7 IE 8	

	Action	Expected Result	Pass/Fail	Actual Result / Issues/Notes
1.	Click the link below to access the SuccessFactors site https://performancemanager4.su ccessfactors.com/TBD	You are brought to a SuccessFactors login screen		
2.	Enter Originator username (*aaaa, bbbb etc.*) and password	Home dashboard appears		
2.1	Click the Recruiting link from the drop down menu	Job Requisitions page appears		
2.2	Click on the Create New hyperlink	Create New Job Requisition page appears		

Figure 5.8: Sample Testing Script with Generic Information

For some testing, such as system integration testing, more specific information needs to be captured in the test cases to verify that all possible scenarios are covered. For example, in an Employee Central implementation, there may be several test cases needed for terminating an employee since there can be multiple reasons associated with termination. Each scenario, such as retirement, poor performance, or accepting another position, should be a specific test case. For this type of comprehensive testing, multiple testing scenarios should be created and managed in a test workbook with each worksheet containing the items to test for each event category. A more comprehensive template to capture this information is shown in Figure 5.9.

Tester Name		Role	Hiring Manager	Test Date:	
Actions to be Taken		Term an employee and verify that date term dates are being updated in the payroll system corrrectly. Also verify that all termination actions are being applied based on the termed code.			

Test Scenario	Emplid #1	Emplid #2	Emplid #3	Pass / Fail	Comments
New starter perm US					
New starter perm UK					
New starter Temp					
Change of address					
Change of address US					
Change of name					
Termination					
Termination - Automatic End Dating of General Deductions					
Retirement					
Retirement - Automatic End Dating of General					

Figure 5.9: Test Scenario Summary Template

Production Migration and Validation

At this stage of the implementation, you are a few weeks away from go-live. You have completed UAT and provided signoff to your consultant. While UAT was in progress, the training team has been wrapping up the training documentation and the change management team has been finalizing the communications. At the same time your consultants have been working on the cutover checklist. A sample copy of the checklist is shown in Figure 5.10. This checklist is a critical part of the migration and validation process. It contains standard activities that apply to all implementations, and it should also reflect your unique requirements.

Production Validation and Go Live Check list

Group	Task No.	Category/Task	Due Date	Completed Date	Status	Owner	Notes
All Groups	1	Period			Open	3DR/Client	
RCM	1.1	Copy/Screen Print Company Settings in Production Provisioning			Open	3DR	
EC	2	Objects to enable Employee Central, MDF and RBP			Open	3DR	
EC	3	Grant permission to Global System Super Admin to login / revoke login access to all users.			Open	3DR	
RCM	3.1	Set Up Provisioning/ Company settings			Open	3DR	
RCM	3.1	Set Up Provisioning/ Company settings Recruiting			Open	3DR	
RCM	3.1	Set up Provisioing / Managing Recruiting			Open	3DR	
Onboarding	3.1	Copy QA ONB site to Production			Open	Development	
Onboarding	3.1	Create FTP folder and login credentials			Open	Development	
Onboarding	3.1	Enable Document Center			Open	Development	
Onboarding	3.1	Check SAML Audience and other are set for production			Open	Development	
Onboarding	3.1	Update BizX URL's for Onboarding			Open	Project	

Figure 5.10: Go-Live Checklist

Once the checklist has been reviewed and agreed upon and the dates finalized, your consultants migrate your configured system to the production instance. If this is the first phase of your implementation, the production instance will be an empty shell. It is important that all changes (data or configuration) to the production instance are controlled, since once a change is made to the production instance it is difficult to remove. It is not the configuration itself that is difficult to remove; it is more that user data associated with the configuration will be affected. For example, if a custom department filter field is used for reporting and chargeback and then the filter field is removed, your reporting and financial process will be affected since the information is no longer in the system.

To start the migration, your consultants will first enable the appropriate features and modules in the production instance. They will copy over the configuration and templates. They will work with you to ensure all the picklists, role-based permissions, platform components, and foundation data are updated in the production instance. At this point, actual employee and talent management data can be loaded into the instance. Your consultants will spot check the system and then turn it over to your team for production validation.

When doing the production validation, work with your consultant to confirm what data to use. For some modules, live employees can be used, while in other cases, such as performance reviews, you do not want to create test performance forms for live employees because they would become a permanent part of the employee record.

The validation process is not meant to be another UAT. It is to validate that the configuration that existed in the test instance is now reflected in the production instance. At this point, if there is a configuration change because of a missed requirement, then that is different and will be tracked as part of the production go-live open items.

For large implementations with several modules and integrations, it is best to have a daily review call with your implementation partner during the production validation period to quickly address any open issues. It is also helpful to separate the critical issues from the nice-to-have items. One definition of a critical issue generally agreed upon by both clients and implementation partners is that it is an issue that will prevent the system launch. You can use a template like the one in Figure 5.11 to provide a summary of these issues. A similar template can be used for both UAT and go-live.

UAT Testing Summary						
		Critical Issues		Non Critical Issues		Closed
Module/Category	Total Test Scripts	Open	Retest	Open	Retest	Closed
Integration						
Goal Management						
Performance Management						
Employee Central						

Figure 5.11: Summary of Open Issues

This summary consolidates all the open issues and enables you to quickly determine how many are critical for which module. You can focus on the critical issues related to order in which you're launching the modules.

Once all the critical issues have been addressed and you are ready to launch the system, the consultant will request a production sign-off. This is to confirm that the production system has been configured based on your design requirements and it has been fully validated by your team. This artifact is also needed by SAP SuccessFactors so they are aware that your system is live and support will be transitioning from the implementation partner to the SAP support team.

Chapter 6

Post-Implementation Planning

There are several areas that need to be addressed when considering post-implementation planning. In the first part of this section, I will look at system support—both the longer-term, ongoing support model and the more immediate support transition process that is done at go-live. Later, I will describe a post-review analysis of the project, which is meant to highlight areas needing improvement during the next phase of the project.

Post-Implementation Support

From using other systems, you know that a knowledgeable and well-trained support team is critical for both the successful rollout and ongoing maintenance of the system. The same is true for SAP SuccessFactors even though it is a cloud-based system. It is key that the support process and the support team be defined and trained before your system launch date.

This expanded team will likely consist of internal employees, your SAP SuccessFactors support team, and representatives from your implementation partner and third-party vendors. This was why we recommended including post go-live support capabilities as one of the evaluation criteria when deciding on an implementation partner.

One approach to defining how post-implementation support will be done is to create a governance body and assign them this responsibility. This approach ensures that support is not a one-off reactive process focusing only on current issues and bugs but instead covers the overall health of the SAP SuccessFactors system and provides a guiding strategy for how the system will evolve.

Typically, the internal team will include representatives from your operations, human resources, and/or information technology departments, and they will be responsible for implementing the tools, processes, and mechanisms to address current support issues and future needs and requirements. I will cover future needs, quarterly releases, and enhancements in the next chapter. For now, your internal team should focus on how support is currently being provided for other systems. Here are some of the key items for them to review:

- How does the organization support technology systems today?

- Is there a help desk?

- What kinds of questions do they typically solve?

  ○ Does the functional owner own support?

  ○ What does the support and escalation process look like?

  ○ Are there any service level agreements (SLAs)?

- Can the existing support model, or parts of it, be leveraged for the HCM suite?

- What type of training is needed for our support team if they can be used for tier 1 support?

With answers to these questions, your internal team can evaluate if the existing support model will work well for the SAP SuccessFactors system. Since this system will be used by all employees, you may need to make some changes to the support model. One potential change is to provide some self-service support capabilities to the user in keeping with the self-service nature of the system. The Jam tool or other in-house collaboration tool, such as Microsoft SharePoint, can be used to store information on how to address common issues. The key is to consider the user as part of the overall support model. Figure 6.1 shows a diagram on how the user can be the first level of the support process.

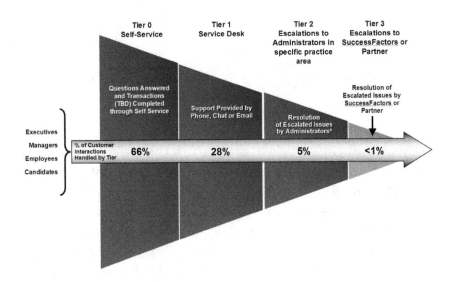

Figure 6.1 User Engagement in the Support Process

The next step is to define how the actual support process will work. It is helpful to have this in a graphical format so it can be included in the training guides with the appropriate contact information. To do this, you will need to define the activities that will occur at each step in the process, who will be responsible for this support activity, and how the support team will be trained. Include a task to ensure that the support documentation can be updated and improved as the team becomes familiar with the system and are exposed to new information and lessons learned. Figure 6.2 shows a typical support process.

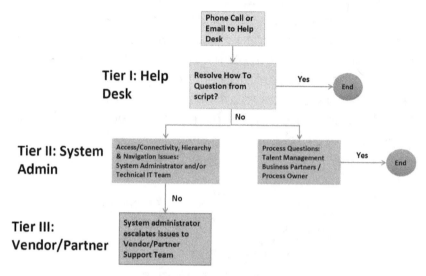

Figure 6.2: The Support Process

A key role in the support process is the primary system administrator. I will cover this role in more detail in the support transition section, but it is important that the right person be selected for this role. Organizations often do not perform due diligence when selecting the primary or secondary system admin. While the support responsibilities are being transitioned, this won't be as obvious, but when it is time to plan and execute changes for the succeeding year it will become more noticeable.

SAP Support

If you realize there is no qualified person internally to be your system admin, then you still have options. SAP SuccessFactors offers different support models. You can use the SAP Preferred Care for more personalized support to supplement SAP Enterprise Support. SAP Preferred Care includes everything in SAP Enterprise Support, but SAP Preferred Care customers receive:

- priority handling and enhanced service levels for response and resolution targets;

- an assigned customer success manager (CSM) with functional and technical expertise on the solution;

- weekly case reviews with assistance on escalations, when necessary;

- access to individualized system administrator training plans;

- cycle, go-live, and integration support;

- individual guidance by product experts, when needed;

- guidance for release optimization and testing best practices;

- support governance and quarterly scorecards;

- access to technical services to do minor configuration; and

- assistance with refreshes of test instances.

Partner Support

The other option is to retain your implementation partner to support the system. Depending on the type of implementation, you may still need your implementation partner for support until your internal team can be trained. The reason is that while SuccessFactors will provide support for the system and the integration platform, making modifications to existing configurations, custom integrations, data issues, and reporting are not covered.

Another benefit of retaining your implementation partner for support is that they can troubleshoot and resolve issues more quickly than the SAP SuccessFactors support team who will not be as familiar with your configuration. In any case, if the support is related to modifications to existing configuration, custom integrations, data issues, and reporting, SAP support will request that you contact your implementation partner to resolve these issues.

If you decide to use an implementation partner to provide additional support, then clarify the areas that they will support and update the support documentation discussed in earlier in this chapter so your admins know who to contact. Here are some areas to consider for your implementation partner to support:

- managing system elements, including performance and recruiting forms, goal plans, employee and candidate profiles, translations, and platform elements like themes;

- preparing for compensation cycles including form management, data preparation, and support;

- building and supporting calibration sessions for performance, compensation, and succession

- providing support, guidance, and training to administrators;

- monitoring nightly file feeds, research/triage, and working with IT and SAP SuccessFactors to resolve any issues;

- working directly with IT on technical issues relating to SAP SuccessFactors;

- evaluating SAP SuccessFactors quarterly releases for system impacts and making recommendations for upgrades;

- troubleshooting and resolving issues;

- documenting and maintaining system administration processes and procedures; and

- supporting dashboards and reporting needs.

The support capabilities for both SAP SuccessFactors and your implementation partner should be reviewed and compared with the level of support that your organization needs. Standard support from SAP SuccessFactors is included at no additional cost with the software licenses. SAP Preferred Care or use of partner personnel for support will incur additional costs, so this is something you will need to review early in the project. This way you can budget for these additional costs, define how the support process will work, and determine the key roles needed.

Support Roles

To define the roles and responsibilities of the support team, you may want to develop a chart that indicates which roles are responsible, accountable, consulted, and informed (RACI). Figure 6.3 shows a sample chart that you can use as a starting point to create your own version. This example is an initial request from a user, so primary responsibility falls on tier 1 support. Tier 2, 3, and SAP, partner, or third-party vendor are all listed as secondary support. If tier 1 cannot resolve the issue, then a support ticket is created and assigned to tier 2. If sufficient information has been gathered to conclude that this is a critical technical issue, then it can be assigned to SAP, the partner, or the third-party vendor.

		P = Primary S = Secondary									
Support Type	Activity / Responsibility	Tier 1	Tier 2	Tier 3	SAP or Partner	What info should be gathered?	Is a script needed?	Who will develop the Script?	What system admin access is needed?	What system training is needed?	Escalation Path / Communication Plan
Troubleshooting	Solve end user issues	P	S	S	SAP	End user details, issue	Yes	Tier 2	Prod, limited admin functions	Tier 1 Training	Create a service request if end user cannot be resolved by Tier 1

Figure 6.3: RACI Chart

You should have a clear understanding of the support process along with the roles and responsibilities of the support team. The final step is to provide the support team with training before the system is launched. Some end-user training can be used here, but more details and exposure to the system are needed for the support team to do their job.

Additional training can be sourced from several areas. SAP SuccessFactors offers admin training, which provides a more detailed overview of admin functionalities. Another source is your implementation partner. Since the latter is very familiar with your configuration, they can assist with the creation and dissemination of the support training if they have the appropriate personnel. Even if this is a not a service they provide, you can still request some knowledge transfer sessions so your internal team can become more familiar with the configuration.

You can also engage with a third-party vendor that specializes in developing training materials. As an example, frequently asked questions (FAQ) documents are very effective for both end users and tier 1 support. Figure 6.4 shows an example of different types of FAQs.

Tier I FAQs

Typical questions can be grouped into three categories: "TECHNICAL", "NAVIGATION", and "PROCESS".

TECHNICAL

How do I log into SuccessFactors?
When I click on the SuccessFactors link, I get redirected to the COMPANY login page
The home page is not loading; the system times out or is very slow
Can I access the system from home?
Do I have to be connected to the internet to see my objectives?
I can't type content into the objective plan fields
When I click the user guide link on the objective plan or in the Welcome portal on the home page, nothing is opening.
The org chart isn't loading
My personal information is incorrect, how do I update it?
My manager is wrong in the system
The system shows an incorrect direct report
I can't type information into the editable fields in my profile
The drop downs don't work in my profile
I entered information into my profile but it won't save

NAVIGATION

Is there a user guide?
How do I get to my objective plan?
Can I print the org chart?
Should I have something in the Performance Tab?
When I click the user guide link on the objective plan or in the Welcome portal on the home page, nothing is opening.
How do I print my objective plan?
Some fields are missing on my objective plan.
How do I see my other sales objective plans?
How do I see my direct report's objectives?

Figure 6.4: Sample FAQ by Category

Support Transition

Once the system has launched, the primary objective for the next two weeks is to ensure that you have adequate support during the initial post go-live period. During this period, you will also transition some or all of the support over to SAP SuccessFactors. In this chapter, I will look at this transition process.

The support transition process varies based on the SAP SuccessFactors support model you have selected. If you've chosen SAP Preferred Care, the support representative assigned to your organization early in the project will provide the needed support during go-live. If you've chosen SAP Enterprise Support, then I recommend continuing to use your implementation partner for all support activities during the immediate post go-live period. You may incur additional costs for this, but it is well worth it as the system is being used for the first time by the majority of your employees. It is important that any issues be resolved quickly and that you get knowledgeable and timely support during this period. This is particularly true if you are doing a big bang rollout instead of a soft launch.

You can work on the transition during the go-live period, but I recommend waiting until you are fully live before completing the support transition. Until your own team is very comfortable with the system, it's better if your implementation partner continues to support you.

The Admin Role

A key internal role in the support transition process is that of primary system administrator. While you may have multiple administrators for various modules, the primary system administrator owns the shared platform components and serves as the main point of contact with your SAP SuccessFactors support team. All support requests need to be routed through this person (or their designated backup) to SAP support.

Here is a more comprehensive list of items that can fall under the primary administrator's responsibilities:

- manage and escalate any application issues

- configure and implement system features

- customize the environment to reflect the organization's business processes

- start processes and initiate activities at key performance milestones

- update the security model and make changes to user permissions and privileges

- update picklists, foundation, and organization information

- create, permission, and deploy reports

- serve as the primary contact between users of the system (primarily via HR representatives) and other internal systems (i.e., PeopleSoft, SAP)

- serve as the primary contact between users and SAP SuccessFactors

- manage/oversee user information (data imports, passwords, notifications)

- implement, communicate, and train users on new functionality

As you can see, the SAP SuccessFactors primary administrator role is critical to the ongoing success of your system. When identifying the right person for this role, look for someone with SAP SuccessFactors system expertise and knowledge of your unique business processes and configuration. They will need to maintain their knowledge of the SAP SuccessFactors system by keeping pace with system upgrades and configuration changes to your system. Ensure that the backup administrator is similarly qualified so if the primary administrator is out on vacation, or worse, leaves the organization, then the backup can step in.

Although I show the primary administrator as a main point of contact, I recommend that during the transition period, the project manager still continue to own and manage any open issues. After the support transition has occurred and the project manager has cycled off the project, then the primary administrator can assume ownership of all points of contact.

Training the System Administrator

Clients often ask how they can get the primary and secondary admins trained so they can fill this role. The short answer is to get them involved very early in the project and be clear about their eventual responsibilities. They should work closely with the implementation team and even own some of the project activities such as data loads, picklist updates, and user permissions.

Other than being actively involved in the implementation, the system admins should attend several training and knowledge transfer sessions. Some of these will be delivered by your implementation partner at different stages in the project. SAP SuccessFactors also offers training sessions for system administrators.

In addition to the training from SAP SuccessFactors and your partner, there are some websites that have information geared towards admins. I recommend frequent visits to:

- The Customer Community site

- Admin Resource Page in the SAP SuccessFactors instance

- SAP Support Portal

Figure 6.5 shows support menu options in the Customer Community site. Next to the Support Menu is the Training option. You must be an SAP customer to gain access to this site.

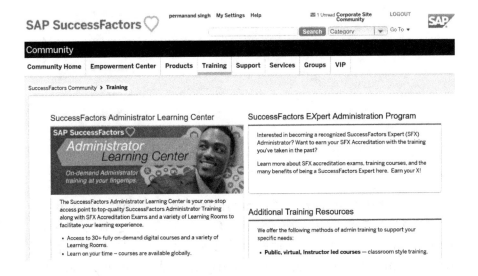

Figure 6.5: The Customer Community Site

The Products menu option contains information on quarterly product updates. Figure 6.6 shows some of the options from this section. On the right side are the archives so you can access information on past releases.

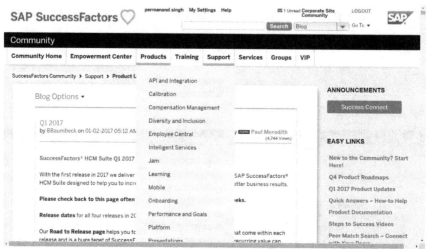

Figure 6.6: Product and Quarterly Release Information

Also on the Customer Community site, there is a new section called Empowerment Center. There are many tools and documents located here that would be helpful for an admin and your project manager to review before and during the implementation. This section is interactive and contains step-by-step details to guide you on the implementation process.

Another source of information is the Admin Resource Page found within the SAP SuccessFactors instance. Once you are logged in, selected the Admin Tools menu option from the home page, verify that you are using the OneAdmin option, and then scroll down to Resources and Materials as shown in Figure 6.7.

Figure 6.7: Accessing Help Links in the System

From here you can get access to individual help topics and also a listing of other key sources of information. Figure 6.8 shows a screenshot of this page.

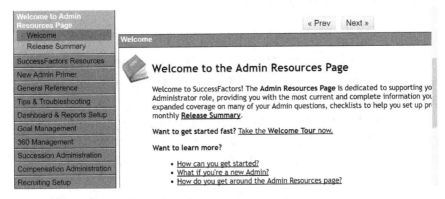

Figure 6.8: The Admin Resource Page

If you are using the new Admin Tools option called NextGen Admin Center, then you can use the Support option to search the knowledge base. Figure 6.9 shows this feature.

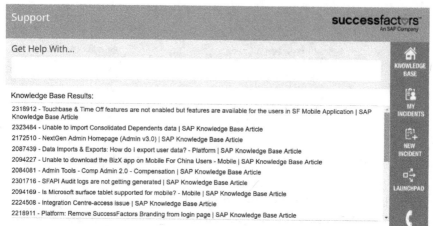

Figure 6.9: NextGen Admin Support Knowledge Base

The SAP Support Portal

The SAP Support Portal is where you will submit support requests and search their knowledge base. Previously, access to this system used to be granted as part of the go-live transition, but it is now occurring much earlier in the implementation process. This is beneficial, since your admin will already know how to use the system and submit support requests before a critical issue arises.

Once the implementation has started and your instance has been created, SAP will request that you provide the contact info for your primary system administrator. SAP will create an S-ID account for this person and give him or her the appropriate privileges so he or she can create the other user accounts for both your internal team and also for members of the partner team. The partner's consultants will need this access to submit any support tickets until the support is transitioned over to SAP SuccessFactors.

The SAP Support Portal is a comprehensive system with extensive features and information. Your admin should get assistance from your implementation partner before submitting their first ticket.

Post-Review Analysis

The post-review analysis is one of the final deliverables for the project and is used by the project manager and senior level management to assess the success of the project, identify best project practices, problem areas, and provide suggestions for improvement on future projects. If you have a phased implementation, then the output from the post-review analysis should also be used during the planning stage for the next phase.

To conduct the post-review analysis, it is best to define the objective of the analysis and create a template to capture the information. This template can be circulated along with the appropriate communication to the key team members for their input. This information is then consolidated into a master copy that is reviewed and finalized during some internal meetings. Depending on your organization, this process may require multiple meetings and is typically managed by your implementation project manager in order to agree on the lessons learned, opportunities, and follow-up actions.

Here are some sample objectives of the post-review analysis:

- review and validate the deliverables and success of the project

- identify project highlights and accomplishments for future projects

- identify problem areas and how problems were mitigated/dealt with

- outline key lessons learned/key takeaways from the project to apply to future projects

This should be an open forum where participants can freely communicate about any issues without being concerned that it will reflect negatively on them. It is an objective analysis of what worked well and what did not. As such, it is a good idea to set some ground rules. One of them is that the scope should also not be restricted to the partner or the vendor but should also consider the internal areas of responsibility.

To define the template to capture this information, some thought should be given to categories to use. Some examples are shown in the phase/module column in the sample template in Figure 6.9. It can be by project phases, by module, by department, or it can be a combination. It is always helpful to provide a sample of the type of information expected and still be flexible if a team member wants to add an additional category or provide additional details.

Successes			
Project Phase/Modules	**What went well/Why?**		
Project Management	The number of team members and the delineation of roles and responsibilities was clear and well planned; good representation across the business. Everyone who needed to be a part of the creative process was invited and welcomed, and input on the design specifications and creation of deliverables was provided by the important project players.		
Design Specifications			
Testing			

Improvements		
Project Phase/Modules	**What Didn't Work**	**Considerations for Future Successes**
Project Management (Timeline)	Timeline was too tight and didn't allow for hiccups; project plan could have been more detailed.	Establish a more detailed project plan and build in extra time to allow for setbacks in the future. Need to allow more time for the unexpected things that come up over the course of a project.
Reporting		

Figure 6.10: Post-Analysis Review Template

Planning for the Next Phase

In a multiphase project, the project team should complete a transition with the team members working on the next phase. It is always very helpful if there is some continuity in the team. If you can retain the project manager, admin, and technical resources from both internal and partner teams, this will ease the transition to the next phase.

Using a knowledge collaboration tool such as SharePoint or Jam or even a shared network folder with all the phase one project documents will help the new team retrieve this information as needed.

This transition process can also be formalized. If your organization has a PMO they may have their own transition checklist and a formal meeting to pass the baton to the new team. They may also have specific documents that the prior team must complete and share with the team working on the next phase. Be sure they share the results of the post-review analysis.

To realize the full benefit from post-review analysis, follow-up actions should be defined and incorporated into the next phase or future projects. Figure 6.10 shows a sample spreadsheet that can be used to track lessons learned and assign responsibility to ensure that follow-up actions are completed.

Lessons Learned					
Module/Activity	Description	Action	Responsible	Status	Notes
Project Management	The effort for the Project Management role was substantially more than initial estimates and could not be done by just one full time employee	Use multiple Project Managers. Have one overall Program Manager and then use a separate Technical Project manager and a separate Functional Project	Bill Effron	Completed	Resource estimates for the Project Manager Role has been increased for the next phase
Testing	Testing took longer than anticipated	Include sufficient time in the project plan for the next phase	Jenny Leu	Not Started	Retesting of some features were needed as they were cross module dependencies
Integration	Our internal integration team needed more time and support to build out the integrations	Get representation from all impacted technical teams from very early in the project and have them contribute and commit to the project	George Will	In Progress	
Recruiting					

Figure 6.11: Lessons Learned Example

Chapter 7

Planning for New Releases and Enhancements

New releases are applied to the system on a quarterly basis, and each year, SAP publishes the release dates for these updates. Your admin should track these dates and not plan any production changes around the same time. If you are currently doing an implementation, it is also recommended that you do not plan any major testing or production rollout during the period when the quarterly release will be applied to your system.

In addition to the release schedule, SAP SuccessFactors also publishes details on each of the releases. Each release will contain changes that fall into two categories: universal and optional. The universal changes are automatically applied to your instance on the scheduled date. For the optional features, you will need to contact your support representative or implementation consultant to enable these features. Then you should test them before they are applied to your production instance.

The application of the universal changes is staggered between test and production instances. Your test instance will receive the update a month earlier than the production system. This means you will have a month to do any regression testing for the universal features and to decide which optional feature to roll out to production.

I previously mentioned that your governance team should determine which new features from the quarterly releases should be rolled out. This team can also be tasked with managing the regression testing and communicating the blackout periods when no changes should be made to production.

Information on the new features will be communicated to customers, but you can also access the information in the Customer Community. The Community also has release archives for access to prior release notes. Finally, SAP SuccessFactors hosts release webinars where the product team covers all the changes coming in the new release. These are an invaluable source of information, and you can often get questions answered on these calls.

Expanding User Adoption and Roll Out

As a former SAP SuccessFactors implementation consultant, I always advised my clients to make their talent management process as simple as possible. The same concept should be applied when deciding on system design and the configurable options and features in the system. Only use features that will augment and support the core objectives of the system.

However, following the initial implementation, I also recommend conducting an annual review to define how the system can evolve. There are several areas to consider when making changes to the system.

1. Unique requirements from distinct user groups: These can occur in several cases, for example, if the initial implementation was for the country where the corporate office is located and now the rollout is being expanded to different countries. Similarly, if the initial target population was a specific division or level, then the rollout can be expanded to other departments or segments of the employee population. In each of these cases, it is likely that some changes will be needed for the system to support any unique requirements for the expanded audience.

2. Modules that are not currently implemented: This is the time to evaluate whether additional modules

would be suitable candidates for expanding the use of the system.

3. Enhancements from the quarterly releases: While this should fall under the purview of your governance team, an annual review is also an opportunity to discuss implementing enhancements.

4. Features that were not included in the initial implementation: There could be several reasons why an existing feature was not included in the initial implementation. It could have been too foreign to your company, requiring a substantial change management effort, or simply that the process in question did not require the feature. It may now be time to consider it.

5. User requested changes: There may be a few vocal users in the organization that will volunteer opinions about changes they would like to see in the system. Regardless of whether you received unsolicited feedback, it is a good idea to proactively reach out to users with a survey to collect their input.

The annual system review is a good time to examine your process maturity level and define what the next level will look like. For example, if your initial implementation was to automate existing processes, then the next level could be to standardize HR processes across the organization and make them more efficient.

There are many HR process maturity models, most of which will have three to five levels. SAP SuccessFactors has a model with 5 levels, as follows:

- Undefined: reactive, lack of integration, lack of transparency, demotivated/disengaged workforce, huge HR admin effort

- Efficient: alignment, consistency, transparency, compliance, efficiency, centralized employee transactions

- Insightful: reduced turnover, data-driven decisions by executives and managers

- Strategic: systematic operations, world-class processes, high-performing organization, executive sponsorship, strategic HR partnership

- Transformational: inspired workforce, clear sense of direction, values-driven organization, change-ready

As you can see, there are many different factors to consider when defining how the systems will evolve. Managing all of these changes will take time and resources, but if it is done well, your SAP SuccessFactors HCM suite can evolve to be one of the most important systems in your organization. In fact, can you think of another system that can play a bigger role in transforming your organization? The effort will be worth it.

Index

About the Author

Permanand Singh has over twenty-five years of experience in technical systems, twenty of which have been heavily involved with HR systems, technology and business processes. Permanand spent four years working for SAP SuccessFactors where he advised on all system and process aspects of an implementation for over thirty local and international clients of varying industries and sizes.

Permanand is a subject matter expert for talent management solutions and SAP SuccessFactors system architecture. He has diversified expertise in multiple facets of the SAP SuccessFactors HCM suite, having led integration projects and several large multimodule implementations. He has also worked as an implementation consultant and implemented most of the modules in the SAP SuccessFactors HCM suite.

Currently, Permanand is the Chief Solution Architect for 3D Results. His focus is on defining the optimal deployment approach and cloud solution architecture for clients.

Prior to his impactful work with SAP SuccessFactors and 3D Results, Permanand implemented the product at WCI Communities and Gartner. At Gartner he was one of the very early adopters of the SAP SuccessFactors HCM suite. In his role as senior director, he functioned as the strategic business partner for the HR, finance, and real estate departments to define business cases and project scopes. Before his time with Gartner, he was the director of HRIS with Ann Taylor Stores, where he managed numerous implementations such as PeopleSoft, benefits systems, ADP Payroll and Oracle Financials.

www.ingramcontent.com/pod-product-compliance
Lightning Source LLC
LaVergne TN
LVHW050150060326
832904LV00003B/90